DECOLONIZING AFRICAN AGRICULTURE

DECOLONIZING AFRICAN AGRICULTURE

Food Security, Agroecology and the Need for Radical Transformation

WILLIAM G. MOSELEY

agenda
publishing

To my late father, Harry H. Moseley, whose curiosity about the world was infectious; and to my spouse, Julia Earl, who has been a faithful friend throughout my 35-year exploration of African food systems

First published in 2024 by Agenda Publishing

Agenda Publishing Limited
PO Box 185
Newcastle upon Tyne
NE20 2DH
www.agendapub.com

ISBN 978-1-78821-589-3

British Library Cataloguing-in-Publication Data
A catalogue record for this book is available from the British Library

Typeset by Newgen Publishing UK
Printed and bound in the UK by 4edge

CONTENTS

PREFACE AND ACKNOWLEDGEMENTS

I first stepped onto the African continent in 1987. I arrived as a Peace Corps agricultural development volunteer and would be based in a small rural community in southern Mali. This intensive two-year experience was foundational in cultivating my respect, curiosity and appreciation for African food systems. I learned a local language, lived with a family, and immersed myself in farming with different households in their fields. It was here, in this small community of Djitoumou, that I became truly enamoured with African agriculture. This passion led to an initial career in international development work, with a focus on agriculture, environment and food security, working for organizations like Save the Children (UK), the World Bank Environment Department and the US Agency for International Development. These experiences eventually fostered more questions than answers, evolving into a second career as an academic geographer.

Over more than three and a half decades, I was able to revisit the Djitoumou area every two to three years. This allowed me to see what the long arc of development looks like in a small rural community. I witnessed the impacts of cotton booms and busts, the construction of a gravel road through the village, wet and dry rainfall years, the expansion of the local primary school, a gold rush and the associated outmigration of young people, the spread of more conservative Islamic beliefs, the diffusion of cell phone technology, the rise and fall of procedural democracy, and the proliferation of widespread herbicide use. But while many things changed, much remained the same. Most children do not go to school beyond the 4th grade, food insecurity and malnutrition are persistent problems, the majority of young students study by lamp light in the evening, far too many women die in childbirth and agriculture is not profitable in most years. While some families might have a few more material possessions than they did 35 years ago, and are able to communicate better with far flung relatives, many human development measures have not changed, such as access to healthcare, education and nutritious food.

As my academic and research career developed, I was able to spend time with rural and urban communities in other parts of Africa, most notably Burkina Faso, Botswana and South Africa. While each of these national contexts are quite different, and there are clear differences in wealth between Mali, Burkina Faso, Botswana and South Africa, poverty is persistently stubborn and malnutrition remains a major problem. My spouse used to jokingly call Botswana "Mali with money" because of its similar climate, yet dramatically greater wealth due to diamond exports. South Africa also has an export-oriented agriculture sector that some agronomists might hold up as the gold standard to aim for. But if Botswana and South Africa are the future, then donors and African policy-makers need to be asking serious questions about where conventional agricultural development thinking is guiding them. This book suggests that a new road map is in order.

This book would not have been possible without the help and support of many different institutions and people. The text blossomed from the germ of an idea during a sabbatical at the University of Montpellier (in southern France) in the fall semester of 2021. There I was supported by the Montpellier Advanced Knowledge Institute on Transitions, an Institute for Advanced Studies at the University of Montpellier, and housed within the ART-DEV research unit. I was then able to develop many of the ideas in this book further during the 2022–23 academic year when I was a Phi Beta Kappa visiting scholar, traveling to six different universities in the United States to give a series of lectures on related topics and ideas. Much of the actual writing for the book was done in 2023 on my front porch in Saint Paul, MN, USA, where I am the DeWitt Wallace Professor of Geography at Macalester College. A special thanks to Camilla Erskine, my editor at Agenda Publishing, who believed in this project from the start, patiently waited for a good year while I worked out many of my ideas for the text and then gave me invaluable advice on several chapter drafts once I started writing.

I am extremely grateful to a variety of funders who have made my fieldwork possible in the different African countries featured in this book, namely: the Fulbright-Hays Dissertation Program, the University of Georgia and Macalester College for my work in Mali; the National Science Foundation (grant #1539833) and the Womadix Fund in Burkina Faso; Macalester College, the Womadix Fund, and the Associated Colleges of the Midwest in Botswana, and the National Science Foundation (#BCS-0518378), the Fulbright-Hays Program and the American Geographical Society in South Africa. Thanks as well to the Knowledge Unlatched programme for supporting open access for readers of this volume and to the Womadix foundation for covering cartography and indexing costs.

I am also appreciative of the many Macalester College student collaborators whose work I have cited in this book and/or who have been involved in field research with me in Burkina Faso, Botswana or South Africa, including Rachel

Fehr, Megan Grinde, Saemi Ledermann, Julia Morgan, Miki Palchick, Stephen Peyton, Millie Varley, Nick Salvato, Eliza Pessereau, Jane Servin and Zoe Tkaczyk. I am also grateful to local research collaborators in each of the four countries featured in this book where I have worked, including: Lamine Dierra, Zana Sanogo, Solomani Doumbia, Issiaka Coulibaly, Chita Sanogo and Drissa Sanogo in Mali; Adama Ouedraogo, Melanie Ouedraogo, Eveline Héma, Yacouba Zi, Salimata Traore and Bureima Kalaga in Burkina Faso; Thando Gwebu, Abrahm Botlhale, Chasha Masabase, and Gabarate Keti in Botswana; and Mike Meadows, Jane Battersby, Maano Ramutsindela and Lizzie Krueger in South Africa. Thanks as well to two student cartographers, Julia Castellano and Kellen Chenoweth, who revised or designed many of the maps and figures in this book, and to Matt Gunther (at the University of Chicago) for developing some of the figure and data tables in Chapter 5.

I also want to acknowledge the many other scholars whose work or collaborations have influenced my thinking, most significantly Michael Watts, Judy Carney, Paul Richards, Piers Blaikie, Sandrine Dury, Larry Becker, Paul Robbins, Ivette Perfecto, Vicky Lawson, Erin Fouberg, Paul Laris, Maano Ramutsindela, Jennifer Clapp, Tom Bassett, Jane Battersby, Brian Uribe-Dowd, Rachel Bezner Kerr, Kefa Otiso, William Cronon, Hanson Nyantakyi-Frimpong, Diana Liverman, Melissa Leach, Dianne Rocheleau, Tor Benjaminsen, Ed Carr, Alice Hovorka, Michael Darkoh, Abdi Samatar, Rick Schroeder, Dick Peet, Lakshman Yapa, Amartya Sen, Herman Daly, Julius Holt, Heidi Gengenbach, Leslie Gray, Ruth Hall, Colin Bundy, Andries Du Toit, Jessie Luna, Carl Jordan, James Sumberg, Matt Schnurr, Brent McCusker, William Munro, Rachel Schurman, Kavita Pandit and Ikubolajeh Logan. A special thanks to my colleagues in the geography department at Macalester College who have been wonderful friends and supporters (Holly Barcus, Eric Carter, Catherine Chang, Xavier Haro-Carrión, Laura Kigin, Dave Lanegran, Ashley Nepp, Laura Smith and Dan Trudeau). I am beyond lucky to work in such as collegial atmosphere.

Hats off to my parents, my mother Ellie Moseley, and my late father Harry Moseley, for consistently and whole heartedly supporting me on a professional journey that was quite different from their own. To my two children, Ben and Ash, who often lived with me while doing fieldwork in Burkina Faso, Botswana and South Africa. Last but not least, I am immensely grateful to my wife, Julia, who has spent time with me in nearly every place I have worked, kept me grounded, cheered me up when I was down, served as a great sounding board for many ideas and been a wonderful friend. She and I met as Peace Corps trainees (which she sometimes refers to as the international dating service for liberals) and we are still talking about agroforestry nearly four decades later.

LIST OF FIGURES AND TABLES

FIGURES

TABLES

PART I

THE BIG PICTURE

1

INTRODUCTION: DECOLONIZING AFRICAN AGRICULTURE

In 1987 I stepped onto the tarmac at the international airport in Bamako, Mali, on a warm June evening. It was my first trip to the African continent and I was a young Peace Corps volunteer. At that time, I had no idea that I would eventually become a scholar of African food systems, returning countless times to various African countries over the next 35 years. Rather, I was just eager to be there, to explore, to learn, to make new friends and to possibly be helpful whenever it made sense. Even then I was sceptical of conventional development efforts, having read, discussed and debated books like E. F. Schumacher's *Small Is Beautiful* (1973) and *The Limits to Growth* by Meadows *et al.* (2018, orig. 1972) during my undergraduate days. However, at the time I was less critical of mainstream thinking on the environment, population and hunger alleviation. I also did not fully understand the neoliberal economic policies that were driving the transformation of the agricultural landscapes and food systems I would encounter.

Following several months of training in the local language, Bambara or Bamanankan, and tropical agricultural practices, I spent the next two years in a small rural village as an agricultural volunteer. Given that the Sahel region had just experienced a major famine in the mid-1980s, I was to work closely with men and women farmers promoting gardening and locally appropriate food crops. I was loosely affiliated with the provincial agricultural authority and my government counterpart was the local agricultural extension agent. I suspect he found my interest in vegetable production quaint as his main focus was on getting local farmers to produce more cotton. When he and I met the other extension agents at our monthly meetings, I could not figure out at first why the only crop we ever discussed was cotton, including endless discussions about hitting cotton quotas. I would subsequently come to understand that Mali had signed agreements with the World Bank to pursue structural adjustment reforms in exchange for loans. Part of these commitments involved an extreme focus on cotton production, a crop for which Mali was deemed to have a comparative advantage. In fact, as a result of these efforts, Mali would become the leading producer of cotton in all of Africa by the 1990s (Moseley & Gray 2008).

About a year into my time in Mali, my fellow volunteers and I had a chance to meet the director of the US Agency for International Development (USAID) for Mali at a gathering in the capital. I remember asking him why, in the aftermath of a major drought and famine, was the US government supporting cotton production, both directly via bilateral assistance and indirectly via the World Bank. This was an industrial crop that could not be eaten, degraded the soil and was causing farmers to become indebted. His response was illuminating. He made three points that I would subsequently hear repeated by Malian civil servants: (1) that Mali needed to produce cotton to meet its debt obligations; (2) that cotton production was key to poverty alleviation in a country like Mali; and (3) that cotton and food crop production were complementary or symbiotic. Cotton, he argued, solved a number of problems, including boosting food production.

Returning to Mali in the 1990s for my graduate master's thesis and PhD dissertation research, both dealing with agriculture and cotton production, I interviewed a large number of civil servants and agricultural extension agents. Many kept sharing the same refrain as if it were memorized: "*Grace à la CMDT, nos funtionnaires sont payees*" [translation from French: because of the government cotton company, our civil servants are paid]. Another phrase I heard over and over again from state agricultural extension agents was that "*kori tigi ye nyo tigi ye*" [translation from Bamanankan: big cotton producers are big sorghum producers]. It was true that cotton helped the government of Mali meet a lot of its financial responsibilities at that time, but it was also leading many farmers to indebtedness rather than out of poverty. Furthermore, as my own research would subsequently show, the celebrated synergies between cotton farming and food crop production were only true for the wealthiest of farmers, with many others becoming increasingly food insecure (Moseley 2005a, 2008b).

How is it that Malian agricultural researchers and extension agents, largely trained in crop science or agronomy, were providing advice to smallholder farmers that was not good for them? The answer, as I will argue in this book, is that the science of agronomy is not apolitical but rather infused with power and politics. In other words, agronomists have focused on developing specific types of crops, with certain characteristics, for the benefit of certain entities and groups of people. We cannot begin to decolonize African agriculture and address food insecurity until we understand this dynamic.

My position as an outsider

Before we dive into the rest of this book, there are at least two aspects of my positionality in relation to this book's topic that need to be acknowledged. First, I am a white male from the Global North writing about food security and agricultural development policy in the African context. I bring a set of life experiences to this topic that are different from someone who was born and raised on the continent. While I lived for a good decade in various African countries (see Preface), I did so as a privileged outsider. My status as an outsider had pros and cons. On the positive side, it often allowed people to open up to me about their trials and tribulations, sharing thoughts and concerns with me that would have been more difficult to share with a neighbour or community member (which is why I have protected the anonymity of everyone who spoke with me throughout the many years of doing research for this book). On the downside, I also realize that there are many aspects of African life that I will never fully comprehend as I have not lived in the shoes of the many people I spoke with.

Second, African policy-makers do not function in a vacuum, but operate in an international policy ecosystem composed of governments, donors, non-governmental organizations (NGOs), international institutions, think tanks and universities, many for whom I have worked. This raises the second major aspect my positionality, my work with the US Peace Corps, World Bank, US Agency for International Development (USAID), Save the Children Fund (UK), International Livestock Research Institute (ILRI), the UN High Level Panel of Experts on Food Security and Nutrition (HLPE) and various universities (Macalester College, University of Minnesota, l'Université de Montpellier, University of Botswana and the University of Cape Town). Clearly my engagement with these organizations has shaped my perspective on the book's topic. As the reader will discover, I am critical of the role that several of these institutions have played in shaping the discourses and programmes that form the African development complex. However, while I am critical, I also understand that these institutions are not monolithic and that hard working and well-meaning people may push problematic approaches because of the way they were educated or due to the constraints under which they are operating. I have been in their shoes and I have made these mistakes (e.g. Moseley 2007d; Moseley & Laris 2008).

Decolonizing African Agriculture

Development geographers Gillian Hart (2001) and Victoria Lawson (2007) make a distinction between what they call "big D" development and "little d" development. Big D development refers to formal development initiatives sponsored by governments, international institutions and NGOs. In contrast, little d development is the economic change that is happening all the time around us. It is not part of a formal development initiative but rather the result of a multitude of actors who are not clearly coordinated under one authority. An example of these two types of development will be presented in Chapter 5 on Burkina Faso.

Here I will examine an Alliance for a Green Revolution in Africa (AGRA) rice commercialization initiative, an example of big D development, which exists alongside the proliferation of herbicides, a process that is not facilitated by one authority but multiple actors and forces operating at different scales (little d development).

I would not be the first to suggest that big D development is inflected with politics. In his well-known book *The Anti-Politics Machine: "Development", Depoliticization, and Bureaucratic Power in Lesotho*, James Ferguson (1994) argues that a technocratic development business has tried to present itself as apolitical, when nothing could be further from the truth. As is the case for the broader realm of development, agricultural development has also been presented as science-based, technocratic and most certainly apolitical. Innumerable agricultural development efforts were launched by European powers in the colonial period in an effort to transform African agricultural landscapes and food systems. Similar efforts continued in the postcolonial period, often cast as initiatives to develop the country or better feed its population. Agronomists were often at the helm of these apolitical, technocratic efforts, using their scientific authority to design the initiatives and prescribe certain approaches.

While shrouded in the methods of science, agronomists favoured some types of understanding, crops, systems and characteristics over others, even when the evidence pointed in the opposite direction. A good example of this were the practices and approaches privileged during the West African cocoa boom in the early twentieth century. As documented and analysed by Ross (2014), the modern plantation approach was favoured in this period (meaning outside capital and technology harnessed for the intensive production of one crop), even when local African techniques proved more efficient. Ross wrote: "the huge increase in output during this period came overwhelmingly from an expansion from smallholder-style production – by which I mean land-extensive and mixed-cropping techniques" (2014: 51). Yet, in spite of evidence to the contrary, agronomists and colonial officials continued to favour intensive production techniques on European-run plantations. Ross argues that part of the explanation for this denial of reality is about culture and ideas, more specifically, Europeans' "powerful ideological devotion to plantation agriculture and the methods associated with it ... [because] plantations were much more than commercial enterprises; they were also incarnations of European agronomic knowledge and symbols of European power" (2014: 51).

In light of a history of agricultural initiatives gone bad, many have given up on the intellectual project we call development (Hancock 1989; Moyo 2009). I have a different take. For me, simply doing nothing is also a choice that does not avoid problematic outcomes. Even if big D development initiatives were to stop tomorrow, little d development or economic change would continue in ways that are not helpful for many African people. Recently, for example, I heard from a friend in Mali that communities near one of my old research villages now have an active, private land market (a new phenomenon). This is an area where farmers and community members historically could not buy and sell land as they operated under a common property regime,[1] a system where one could only hold usufruct, or use, rights to the land, but not buy and sell it. The privatization of land in Africa's rural areas is a particular problem for women farmers as it means their male relatives, who often hold the use rights to the land, may sell it off to meet short-term needs and undercut the long-term subsistence of the household (Becker 2013). As such, regardless of what you think about big D development efforts in Africa, more generalized and equally problematic economic changes are happening on the continent that undermine the livelihoods of the most marginal and food insecure people. Furthermore, while the political nature of big D agricultural development is well documented, I would also argue that little d agricultural development is shaped by power and politics. For example, with regards to the earlier example, the idea of land privatization has been pushed by a variety of thinkers globally as critical for capitalistic development on the continent (Hardin 1968; Page 2012). These arguments have influenced land codes in certain African countries, thereby fostering the problematic situation of land privatization that hurts women farmers as described above. Politics and power are everywhere, from big D agricultural development initiatives, to the little d economic changes that are happening all around us.

Rather than jettisoning development, I see a need to radically reimagine the paradigms that guide agricultural initiatives and to revive the idea that public institutions are actually capable of doing good. This is not a naïve pronouncement or a call to stick one's head in the sand. Rather, I believe one can do two things at once. One can continue to think critically about development but also conceive of different models and acknowledge the ability of public institutions to change people's lives and food security for the better. As I will argue in subsequent chapters, I see agroecology as a more decolonial paradigm for reimagining African agricultural development that is quite different than previous models that have shaped thinking and approaches in this arena.

Agroecology is an emerging science and social movement. As a science, agroecologists treat the farm field as an ecosystem, studying the ecological interactions between different crops, crops and insects, crops and animals, as well as

---

1. In many rural African communities, land is held in common, and the allocation of rights to use the land is managed by the chief. In most cases, if a male has longstanding roots in the community, and he or his forebearers cleared the land for farming, then he holds the usufruct or use rights to the land. However, he cannot sell the land, but must pass it on to his children. Should the family leave or cease to exist, then the land for which he has use rights reverts to the community.

crop, climate and soil interactions (Carroll *et al.* 1990). By understanding and leveraging synergies within these systems, the farmer may improve production, minimize insect damage and maintain soil fertility. As agroecology allows the farmer to produce more with less expensive inputs, this approach is more accessible than high external input agriculture to the poorest of the poor, the people that are really hungry. Furthermore, agroecology is a more decolonial science as its innovators frequently work across the formal–informal knowledge divide with trained scientists often building on and experimenting with crop combinations and techniques originally developed by farmers and their experiential knowledge (Nyantakyi-Frimpong *et al.* 2017).

While I initially learned about agroecology as a science when I was a graduate student in the early 1990s, agroecology is also a social movement that has close links to food sovereignty. This is important for two reasons. First, given the insights of political ecology and political agronomy (to be discussed in Chapter 3), it is increasingly understood that the agricultural sciences do not exist in a vacuum but are impacted by broader-scale political economic conditions. As such, while it might be rationale to pursue the science of agroecology in order to develop healthier, more equitable and sustainable food systems, social movements are needed to reshape political economy in a way that will allow this new science to more fully emerge. Second, we cannot just graft agroecology onto the existing corporate farming structure as a new set of practices, but the social organization of agriculture also has to be considered. While agroecology existed at the margins of science and policymaking for several decades, it has recently gained traction as a more mainstream discourse in the United Nations system (Loconto & Fouillieux 2019; HLPE 2019).

Many in my academic fields of geography, political ecology, African studies and agrarian studies are sceptical of the ability of large public institutions to do good.[2] Furthermore, neoliberalism has gutted many people's faith in government, especially in African governments.[3] But governments can do good if they are democratic and under the checks and balances of an active civil society. While these are not perfect cases, Botswana under Seretse Khama, Burkina Faso under Thomas Sankara, Mali under Alpha Oumar Konaré and South Africa under Nelson Mandela, did much to positively transform their economies and improve the lives of their citizens. For example, Seretse Khama, as the first president of an independent Botswana, set his government on a path of good governance and shrewd management of its diamond resources, allowing it to quietly

---

2. Robbins (2004) suggests that political ecologists are inherently sceptical of centralized authority, preferring decentralized and community-based approaches.

3. One thing I have observed in 25 years of teaching at the university is that the vast majority of my students have little belief in the ability of governments to do good.

become a middle-income country (Samatar 1999). While much has been written about the role of the private sector in contemporary African development (e.g. Moyo 2009; Mawdsley 2015), I do not believe we can count on corporations, companies and entrepreneurs to improve the lives of the most marginalized and food insecure (Moseley 2015a). Companies by definition need to turn a profit and there is often little to no money to be made from delivering effective approaches that address food insecurity (Moseley 2012a). Food security, certainly for the poorest of the poor, is what economists call a public good, a service or good that is provided to members of society without profit. This quality of food security is consistent with the Right to Food,[4] part of the UN Declaration on Human Rights. As such, there is a critical and necessary role for the public sector to play in delivering food security.

The basic argument

The basic argument of this book has three parts outlined in Box 1.1 and further elaborated below.

BOX 1.1 THE BASIC ARGUMENT

1. Development organizations and governments will only begin to seriously address food insecurity in Africa when they more fully question the assumption that increased crop production, using high external input agriculture, is the solution, an idea that is central to crop science or agronomy.
2. Agriculture development must be seen as more than the first step in an industrial development process, but as a sustainable livelihood that has value in and of itself.
3. An agroecological approach, combined with good governance, will allow people to have greater control over their food systems, produce healthy food more sustainably, and enhance access to food by the poorest of the poor.

---

4. In 2002, the Special Rapporteur on the Right to Food defined the Right to Food as: "The right to have regular, permanent and unrestricted access, either directly or by means of financial purchases, to quantitatively and qualitatively adequate and sufficient food corresponding to the cultural traditions of the people to which the consumer belongs, and which ensure a physical and mental, individual and collective, fulfilling and dignified life free of fear". (Special Rapporteur on the Right to Food 2012).

First, development organizations and governments will only begin to seriously address food insecurity in Africa when they more fully question the assumption that increased crop production, and high external input agriculture, alone is the solution, an idea that is central to crop science or agronomy. Increasing crop yields, trade and production is the main way that policy-makers have sought to address food insecurity in the African context. This was the focus of the first wave of the Green Revolution in the 1960s and 1970s, trade-based solutions during the neoliberal era from 1980–2006, and the New Green Revolution for Africa and value chain approaches from 2007 to the present (Kansanga *et al.* 2019). This narrow approach is not working.

Second, agriculture development must be seen as more than the first step in an industrial development process but as a sustainable livelihood that has value in and of itself. The commercialization of agriculture has long been seen as a step to manufacturing, both in the West and the East. In the West, this idea was central to modernization theory, Rostow's (1960) stages of economic growth being the classic example (see Figure 1.1). Under Rostow's stages of economic growth, the first step was to commercialize more subsistence-oriented agriculture. This eventually led to enough capital accumulation to invest in and develop a manufacturing sector. In further stages, agriculture would subsequently become a much smaller sector of the economy as manufacturing and then the service sector grew, resulting in economies similar to the ones in the Global North today. In the East, the Chinese have a similar but parallel model known as "firing from the bottom" (see Figure 1.1). According to Gulati and Fan (2007), a distinctive feature of China's "firing from the bottom" strategy was its strong initial focus on agricultural growth. This strategy emphasized starting with reforms in agriculture (decollectivization and privatization in the case of China) before moving to the service and manufacturing sectors.[5] That said, the end goal was similar, a more commercialized agriculture leading to an economy dominated by the manufacturing and service sector (Moseley 2013a). In both cases, what these models ignore is the possibility where agriculture is seen as a sustainable livelihood that has value in and of itself, rather than a stepping stone to something else. Is it really desirable or possible for every country in the world to have largely urban economies where farming is highly industrialized and the purview of a few? Furthermore, even in those countries where agriculture is an increasingly small part of the economy today, such as Botswana, it remains a vital livelihood activity for the poor (Fehr & Moseley 2019). Agriculture needs to be reimagined in contemporary African development discourse as not just in need of intensification,

---

5. What is less acknowledged is that growth in agriculture in China is estimated to have contributed to poverty reduction four times more than growth in manufacturing and services (Ravillion & Chen 2007).

Rostow's stages of economic growth

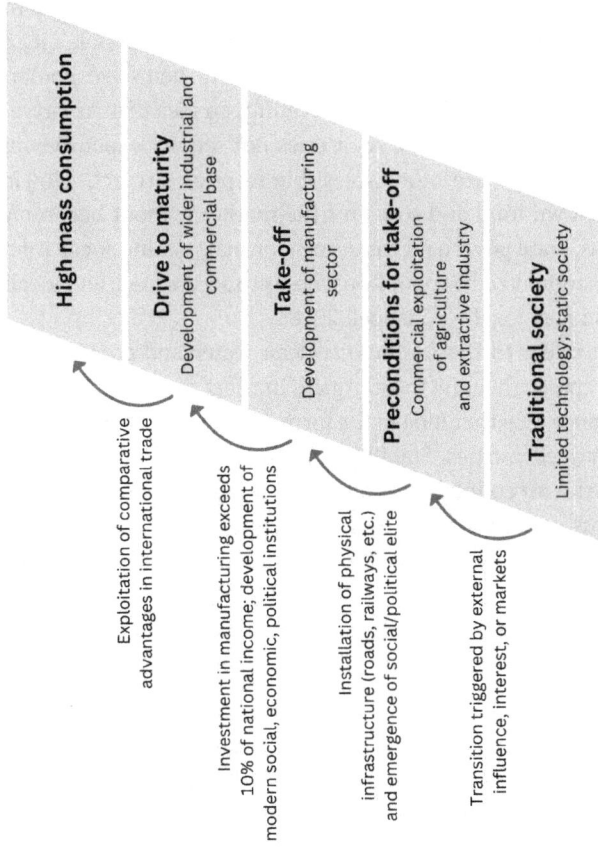

China's firing from the bottom strategy

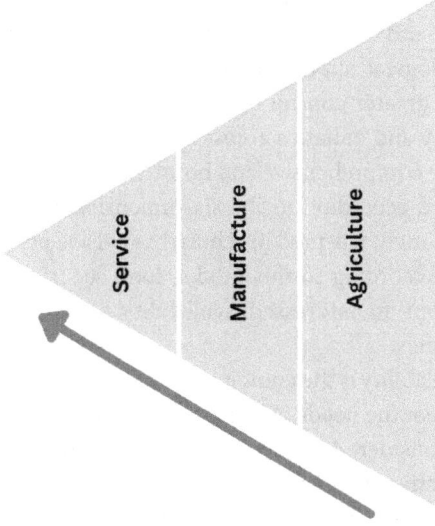

High mass consumption

Drive to maturity
Development of wider industrial and commercial base

Take-off
Development of manufacturing sector

Preconditions for take-off
Commercial exploitation of agriculture and extractive industry

Traditional society
Limited technology; static society

Exploitation of comparative advantages in international trade

Investment in manufacturing exceeds 10% of national income; development of modern social, economic, political institutions

Installation of physical infrastructure (roads, railways, etc.) and emergence of social/political elite

Transition triggered by external influence, interest, or markets

Service

Manufacture

Agriculture

Figure 1.1 The position of agriculture in conventional development thinking: Rostow's stages of economic growth versus China's firing from the bottom strategy

Source: Figure conceived by author, designed by Julia Castellano, Macalester College, and based on Rostow (1960) and Gulati and Fan (2007).

commercialization and concentration, but as a legitimate livelihood that can lift people out of poverty and better feed families and communities.

Third, an agroecological approach, combined with good governance, will allow people to have greater control over their food systems, produce healthy food more sustainably and enhance access to food by the poorest of the poor. This approach is more comprehensive and holistic than high external input agriculture when analysed according to the six-dimensional food security framework (a model increasingly accepted in scholarly, civil society and policy circles) (Clapp *et al.* 2022). According to this model, food security[6] is best promoted when all six dimensions are addressed: availability, access, utilization, stability, sustainability and agency.

Adequate food availability is the combination of local production and net food imports needed to meet the needs of a given population. When all types of outputs and inputs are considered, agroecological approaches have been shown to be as or more productive than conventional approaches (Pretty 2006; Akram-Lodhi 2021). Furthermore, agroecology is more sustainable as it leverages the forces of nature to maintain soil fertility and manage pest and weed populations, rather than chemicals and fossil fuels. Access refers to people's legal means of getting food that is available, either by growing it themselves, buying it on the market or getting it through friends and family, an idea that Amartya Sen (1982) conceptualized as entitlements. As it does not require expensive inputs, agroecological approaches are more accessible to poor farmers, allowing them to produce their own food and some for the market without becoming indebted. In urban areas, good governance is crucial for maintaining social safety nets that allow more marginalized or poor households to access food, and ideally food that has been produced locally (Chappell 2018).

Utilization refers to the sanitation, clean water and cooking infrastructure needed to prepare a healthy meal. Again, in most cases this is closely linked to good governance, whether it be in the form of bore holes in rural areas or proper sanitation services in cities. Stability is consistent food supplies and prices over time. Many African countries became increasingly dependent on food imports with the focus on agricultural trade throughout the neoliberal food security era (1980–2006). This proved to be devastating for those West African countries reliant on rice imports during the global food crisis of 2007–08 (Moseley *et al.* 2010) and those in North Africa and the Horn dependent on wheat imports during the war in Ukraine in 2022–24 (Moseley 2022a). Greater food

---

6. Food security has been defined in a variety of ways over the years, but the most frequently cited formulation today is that it refers to a situation where "All people, at all times, have physical, social and economic access to sufficient, safe and nutritious food that meets their dietary needs and food preferences for an active and healthy life" (HLPE 2020).

production at home via agroecology is needed for a rebalancing of food sourcing and greater stability in food supplies and prices. Lastly, agency, or control over one's food production and consumption choices, is central to an agroecological approach. Farmers have control over this knowledge and methods and grow food for local markets that is culturally appropriate.

Organization of the book

This book is organized into four major parts. In Part I, the first three chapters offer a big picture introduction to the text's major themes and African agricultural development. Following this introduction, Chapter 2 examines a failed history of food security and agricultural development efforts in the African context. Chapter 3 then unpacks the book's five major conceptual frameworks: intellectual decolonization, political ecology, political agronomy, food security and agroecology.

Part II, entitled "Country studies of failed agricultural development in the colonial and postcolonial periods", takes a deep dive into the problematic agricultural development trajectories of four African countries where I have spent most of my career working: Mali, Burkina Faso, Botswana and South Africa (see Figure 1.2 showing where these countries are located). While this is a somewhat random selection of countries, they do conveniently provide two low-income countries (Mali and Burkina Faso) and two middle-income African countries (Botswana and South Africa). They also represent two former French colonies (Mali and Burkina Faso) and two former British colonies (Botswana and South Africa). Lastly, they represent a diversity of agricultural economies. Mali and Burkina Faso are places where smallholder farming really dominates and the majority of the population is rural. In contrast, both Botswana and South Africa are more urban. Animal husbandry dominates in semiarid Botswana and large-scale commercial farming is widespread in South Africa. A final twist is that South Africa was a white settler colony whereas the others were not. More specifically, Chapter 4 examines cotton-based agricultural development in Mali, Chapter 5 looks at the New Green Revolution for Africa and proliferating herbicides in Burkina Faso, Chapter 6 explores Botswana growing wealth, inequality and hunger, and Chapter 7 examines big agriculture's takeover of South Africa's land redistribution programme.

Part III, entitled "Reimagining African food systems", is more uplifting than the second section, examining signs of hope in the same four countries. In some cases, these are deliberate attempts at a different approach. In other cases, the situations were almost accidents or what Turner (1989) might call "natural experiments". Agroecology, the application of ecological principles to agricultural

Figure 1.2 African case study countries
Cartography by Kellen Chenoweth, Macalester College
Sources: Esri Africa 2018; Africa Albers Equal-area Conic Projection.

systems and practices, also features more prominently in this section. Chapter 8 looks at an accidental experiment with food sovereignty and agroecology in Mali during the global food crisis of 2007–08 and the relevance of this to a country wracked with political insecurity in recent years. In Chapter 9, the importance of foraging as a source of dietary diversity in Burkina Faso is highlighted, something encouraged by agroecology's emphasis on thinking beyond the farm field and about broader food systems. In so doing, the chapter highlights the significance of food environments, or the landscapes in and around communities that feed them. In Chapter 10, I examine a backyard gardening programme in Botswana that was launched by the country's former president Ian Khama. While largely seen as a failure, this programme was somewhat unique in its support for women farmers, all the more important in a country where men's animal husbandry activities have long been the focus of agricultural assistance programmes. Lastly, in Chapter 11, I examine some alternatives to the more commercially oriented land reform programme. I explore fair trade as well as smaller farms and gardens with a food security focus as alternatives that may hold more promise than initiatives to have historically marginalized groups run large-scale farms.

Before concluding, the fourth and final section of the book (Part IV) explores the broader scale institutions, both at the regional and international scales, that will be needed to bring about and sustain a radical transformation in the way practitioners and policy-makers approach African agriculture and food security. Just like the international institutions (such as the UN-sponsored Consultative Group on International Agricultural Research (CGIAR) Centers) that were created to support the first Green Revolution, agroecology will need similar types of research support. Lastly, in this section, I also discuss the need for an international food security policy making body that is run on participatory and democratic principles.

In sum, after more than 35 years of working on and studying African agriculture and food security, this book is my way to take a step back and take a look at the big picture. This text has allowed me to think through a series of questions that have preoccupied me throughout my career, questions that were also often of concern to the many of farmers, households and policy-makers who generously shared their time with me during periods of research in Mali, Burkina Faso, Botswana and South Africa. Why is this continent so rich in human ingenuity and natural resources not better fed? What type of thinking has predominantly shaped the way policy-makers and practitioners seek to develop agriculture and address food insecurity in the African context? If these ideas and approaches are not working, what are the alternatives? More concretely, why have some approaches failed in certain places and why? Furthermore, what are signs of hope in those places that could be encouraged as the global community seeks to eliminate hunger over the next decade.[7] And finally, what types of changes at the international level are needed for positive change and progress to play out at the local level? This book is my attempt to answer these questions.

---

7. As per the sustainable development goals (SDGs), SDG 2 calls for zero hunger by 2030.

2

A BRIEF HISTORY OF AFRICAN FOOD SECURITY AND AGRICULTURAL DEVELOPMENT POLICIES

While the largest number of hungry people in the world may be found in South Asia today, the highest incidence of hunger is in Africa South of the Sahara. After several years of progress, hunger has steadily grown since 2014, both globally and on the African continent. Today nearly 60 per cent of Africa's people are experiencing moderate to severe food insecurity, up from around 47 per cent in 2014 (FAO 2021) (see Figure 2.1). Although African rates of hunger are still highest in rural areas, poverty and hunger are growing at alarming rates in African cities as the continent urbanizes (Battersby & Watson 2018). While policy-makers have a history of focusing on acute food insecurity or hunger as the major problem in the African context, there are other aspects of malnutrition, such as micronutrient deficiencies and obesity, that are also growing problems. For example, obesity and related non-infectious diseases such as diabetes and hypertension have steadily increased in the continent's wealthier countries and cities (Hunter-Adams *et al.* 2019) as they go through a nutrition transition that has often accompanied urbanization in other parts of the world. Comprehending the basic history of agricultural development and food security on the continent is important for understanding the contemporary situation.

This chapter examines the history and evolution of African agricultural development and food security policies, from the precolonial period to the present. It emphasizes the tensions between the often-conflicting goals of Africa's smallholder farmers and states. In many cases, African farmers sought to manage risk, control costs and navigate labour bottlenecks while producing food (approaches which were often disparagingly framed as risk aversion), whereas states and regimes were more interested in capturing surplus production. In the postcolonial period, many African states have failed to enhance the food security of their populations because they have been overly focused on commercialization, production and revenues as the main measures of agricultural development success.

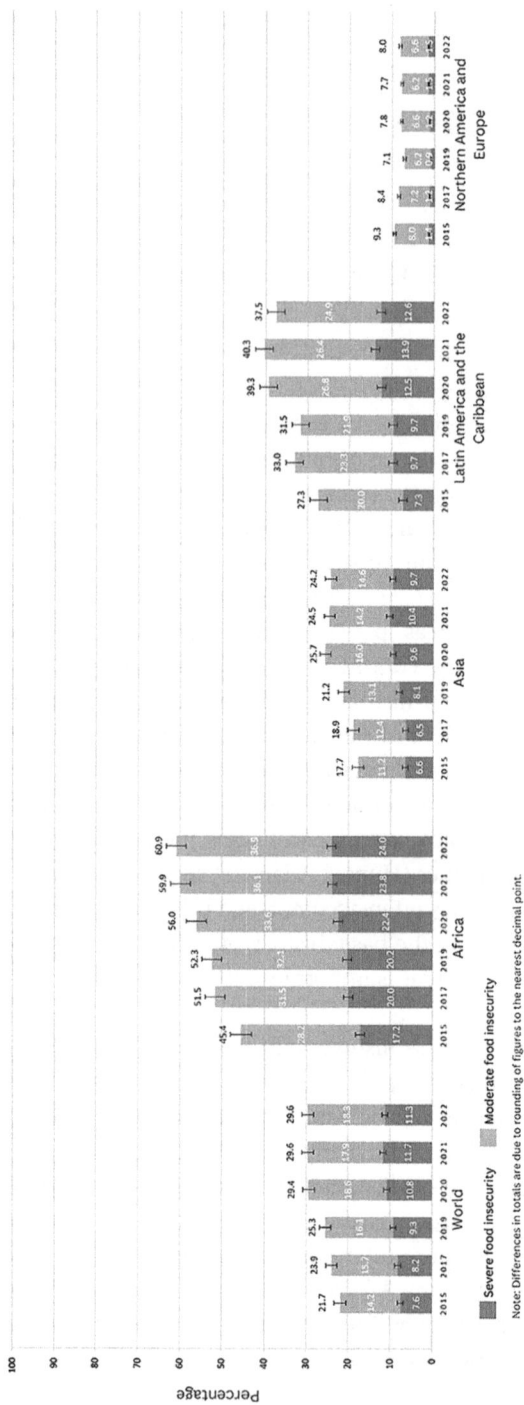

Figure 2.1 Moderate and severe food insecurity in Africa relative to other world regions

Source: Figure by Kellen Chenoweth, Macalester College, and based on information from 2021 FAO SOFI report.

Precolonial African agriculture

In much of Africa, people have long had to deal with highly variable rainfall (Moseley 2008a). This is especially the case in the region's savanna landscapes where crop farming and animal husbandry are most heavily concentrated. Pastoral livelihoods evolved to be highly mobile, with herders moving across vast semi-arid grasslands in search of the best pastures, following the rains as the intertropical convergence zone (ITCZ)[1] moved north and south over the course of a year. Herders also developed reciprocal relationships with farming communities, pasturing their animals on farmer's fields in the dry season (with farmers benefiting from the resulting manure) and trading milk for grain (Moorehead 1991; Davies 2016).

Most farmers grew multiple crops with different rainfall requirements, always seeking to ensure that at least some of the crop would be successful whether it was a lean or overly abundant year for precipitation. The other advantage of polycropping was that it often spread out labour demands with crops maturing in different time periods (helpful since many African farming systems were labour constrained). Rural households also generally stored surplus grain from the good years to get them through the bad rainfall years when little to nothing might be harvested. According to some historical accounts (Franke & Chasin 1980), it was not unusual for an average household in the West African savanna region to have a good one to two years of surplus grain in storage as security in the event of a drought. Furthermore, scholars like Watts (1983) have documented how in Nigeria's Kano region, village councils and precolonial regimes also had a system of grain storage designed to get communities through difficult rainfall years.

Outside of food production, some fibre crops like cotton were grown to supply a local textile industry. The precolonial cotton and textile trade in the Congo is well documented by Thornton (1977), who suggests that in the sixteenth and seventeenth centuries, the region was a major textile producer relative to Europe and Asia at the time (based on his examination of the trading records of Portuguese traders). In West Africa, Roberts (1996) has documented how cotton was frequently grown by women in gardens and supplied to local weavers for the artisanal production of textiles.

---

1. The Intertropical Convergence Zone (ITCZ) is the point of direct solar radiation on the earth's surface. This is at the equator on 21 September, moves south to the Tropic of Capricorn (23.5 degrees S) on 21 December, back to the equator on 21 March, and north to the Tropic of Cancer (23.5 degrees N) on 21 June. This movement is related to the fact that the earth rotates around the sun and is tilted on its axis. The ITCZ creates an area of low pressure and rainfall (related to warm air rising and shedding moisture), leading to seasonal rainfall patterns in tropical regions.

Agricultural change in the European colonial period

Modern European colonialism is different from previous forms of colonialism (such as that of the Greeks and the Romans) as it was about capturing resources to fuel proto-capitalist (read mercantilist) and capitalist expansion, not just securing taxes and tribute (Jarosz 2003). Big picture, scholars often talk about two waves of European colonialism (Fouberg & Moseley 2018). The first wave lasted from roughly 1500 to 1825, with the major powers being Spain, Portugal and the Netherlands (with France and Britain to a lesser extent), and it was largely focused on the Americas. The impact in Africa was most significant in southern Africa with an expanding Cape Colony and along the coasts where several European trading posts were established to facilitate, among other things, the transatlantic slave trade. Most of the interior of the African continent was largely controlled by Africans up until the 1870s, bringing one to the second wave of European colonialism which spans approximately 1825–1975. Here the European actors were Britain, France, the Netherlands, Belgium, Germany, Portugal and Italy (with the first two being most significant).

By the mid to late nineteenth century, there was a fair amount of tumult on the African continent due to depopulation and displacement caused by the transatlantic slave trade. The first Portuguese ships had arrived on the Senegal River in 1444 (Green 2019). While African economies had previously flourished, Green (Green & Thornton 2022) posits that the reorientation of trade away from the East and towards the Atlantic World led to economic decline on the continent, especially as the trans-Atlantic slave trade accelerated in the seventeenth and eighteenth centuries. Lovejoy (1989) argues that massive depopulation occurred in some areas of Africa in this period as a result of slaving. More specifically, Manning (2004) estimates that by the mid-nineteenth century, many of the areas in West and West Central Africa that were human source areas for trans-Atlantic slavers probably had populations that were 50 per cent of what they might have been in the absence of the slave trade. This depopulation combined with encroaching colonial powers resulted in a very unsettled situation in the late nineteenth century with warfare between different African groups and against the colonial forces.

This instability and warfare devastated local food systems. For example, in West Africa, the warlord and resistance figure Samori Touré fought back vigorously against French encroachment in the West African interior in the late nineteenth century, moving back and forth across sections of what is today southern Mali, southwestern Burkina Faso, northern Côte d'Ivoire and northeastern Guinea. In the process, he requisitioned crops and animals in his wars against other Africans and the French, devastating local peasant economies. In fact, in my old Peace Corps village area in southern Mali in the 1980s, my local friends showed me what were sometimes referred to as "Samori holes". These were large underground excavations where entire villages and their livestock sometimes hid when Samori and his army passed through an area, killing farmers and looting crops and livestock (Peterson 2008). Even then, this memory and fear of Samori lingered in some of the minds of the oldest villagers I spoke with. It also led to complicated feelings about the French. In the local language Bamanankan, the word for colonialism is *jonya juru*, which literally means the rope of slavery. People were grateful that the terror of Samori was over, but – at the same time – they essentially felt that they had enslaved themselves to the French in return for protection. The cost was a head tax imposed by the French known as *ni songo* in Bamanankan, or soul tax (again a word with a deeper, weighty meaning). This tax, and the efforts needed to raise the money to pay it, would subsequently have longer term implications for farming systems in the area.

Both the French and British colonial regimes needed to figure out how to extract wealth from African food and farming systems, a process that intensified in the late colonial period as the metropoles of London and Paris became increasingly resource constrained. After experimenting with a variety of approaches, including forced labour, taxation (usually in the form of a head tax) became the preferred strategy (van Beusekom 1990; Bassett 2001). While taxation, or systematic tribute payments, also existed in the precolonial period, the difference was that most of this surplus extraction in the colonial period was leaving Africa. The other advantage of a head tax (from the regime's perspective) was that it created a need for local people to generate cash, a crucial wedge for breaking down subsistence systems. Tropical agronomists and colonial agricultural extension staff were desperate for African farmers to give less priority to their own subsistence and to begin producing surpluses for regional markets and for export (Davis 2002). The need to pay taxes in cash moved this agenda forwards (with local people risking imprisonment if they did not pay such taxes) (see Figure 2.2). The crops grown for sale included some food stuffs, such as maize production to feed the growing urban and mining populations of southern Africa (McCann 2005), but also a wide variety of commodity crops in demand by the colonial metropoles, such as cotton, coffee, tea, tobacco, cacao and sugar (Moseley 2021a).

A key difference between African regions in the colonial period was those with a significant European settler presence and those without. While this is a broad generalization, Europeans tended to settle in those areas with relatively cooler temperatures and lower tropical disease burdens, which meant areas in the temperate zone (poleward of 23.5°N and 23.5°S) and at altitude. More specifically, most European settlers were found in the temperate areas of southern Africa and North Africa and in the highlands of East Africa. In southern Africa under British colonialism, a new regional economy was constructed based on: (1)

CHEFS DE VILLAGE ALLANT REMETTRE LE PRODUIT DE L'IMPOT RECUEILLI PAR EUX. (HAUT-SÉNÉGAL-NIGER.)
CLICHÉ FORTIER.

Figure 2.2 Payment of the head tax in French West Africa
Note: This photograph attributed to Louis Sonolet is entitled "L'Afrique Occidentale Française". The caption with the photo reads "Village chiefs going to deliver the proceeds of the tax collected from their people (Colony of Upper Senegal and Niger)".
Source: New York Public Library, used with permission.

mining; and (2) the production of food to feed the mining and urban popula-tions of the region. Mining was centred around three major poles in south-ern Africa, diamonds near Kimberly, gold near Johannesburg (Witwatersrand Basin), both in what is now South Africa, and eventually copper in the copper belt of Northern Rhodesia (now Zambia). Developing in parallel with the mining industry was a regional maize economy, which would eventually be organized around commercial, European-run farms largely concentrated in the Orange Free State (now South Africa) and the highveld of Southern Rhodesia (now Zimbabwe) (McCann 2005).

The establishment of this new mining and regional maize economy had mul-tiple repercussions for local food economies. First, Africans lost some of their best farmland to white settlers, a topic that will be explored further when we examine land reform efforts in South Africa in Chapter 7. Second, in many cases, Europeans had trouble securing cheap labour and competing with local African farmers, leading to draconian laws that hobbled black farmers (Bundy 1979) and effectively created a system of racialized capitalism. Third, different areas of southern Africa were framed by the British as either productive and in need of labourers, or as labour reserves (less productive areas that would send migrant labourers to other colonies in the region). Two examples of the latter were Malawi and Botswana, then Nyasaland and Bechuanaland respectively. Nyasaland, for example, had roughly a quarter of its male workforce in other countries in the 1950s, working on commercial farms in Zimbabwe (then Southern Rhodesia) and mines in South Africa and Zambia (then Northern Rhodesia). Many of these labourers never returned home, a situation that had longer-term implications for labour-constrained agriculture in Malawi (Moseley 2000). Furthermore, to this day, there are farm workers in Zimbabwe that have Malawian ancestry (Moseley & Logan 2001).

Postcolonial food security and agricultural development policies

Postcolonial food security and agricultural development policies have evolved over time but more or less break down into three eras with an emerging fourth: (1) food self-sufficiency and the first green revolution (1960–79); (2) neoliberal food security and trade (1980–2006); (3) neoproductionism and the New Green Revolution for Africa (2007–19); and global shocks and a growing movement to decolonize African agriculture (2020–present) (see Table 2.1). Herewith I discuss each of these eras.

Food self-sufficiency and the first green revolution (1960–79)

Following a wave of African countries gaining independence in the early 1960s, many countries focused on food self-sufficiency (maximizing food production at home), some cash crop production for a booming commodity trade and limited industrialization. This approach was largely supported by the first Green Revolution and modernization theory. The Green Revolution was a concerted attempt to bring industrial food production techniques to the Global South. It was spear headed with philanthropic support from key American foundations (such as the Rockefeller Foundation), scientific support from agronomists like Norman Borlaug (who would go on to win the Nobel Peace Prize for his work on improved wheat varieties), an American government concerned about growing Soviet influence and a vast array of United Nations institutions[2] concerned about global hunger. While this first wave of the Green Revolution is thought to have largely bypassed Africa, it did have a lasting impact on maize production in southern Africa (Eicher 1995), rice production in some areas of West Africa

---

2. Such as the UN Food and Agriculture Organization (FAO) and the UN Consortium of International Agricultural Research Centers (CGIAR).

Table 2.1 African postcolonial food policy eras

Food policy era	Dates	Pivotal crisis	Policy response	Outcome
Food self-sufficiency and the first green revolution	1960–79	• Global North concerns about population growth • Concerns about global food supply	• Programmes to maximize food production within national borders using inputs/improved seeds • Agricultural subsidies • Public sector led • Develop industries	• Highly uneven impacts across African continent • Increased national level food production in some cases • Agroindustries
Neoliberal food security and trade	1980–2006	• Low and falling commodity prices • High energy prices • Third World debt crisis	• Conditional structural adjustment loans • Focus on commodity exports • Reduction of agricultural subsidies, tariff barriers and government spending	• Increased commodity exports • Increasing reliance on food imports • Deindustrialization • Declines in government extension services for non-commodity crops
Neo-productionism and the new green revolution for Africa (GR4A)	2007–19	• 2007–8 global food crisis • Rise in rice prices • Social unrest in cities	• Integrate smallholder farmers into formal value chains • Engage the private sector in public–private partnerships • Encourage farmers to purchase inputs	• Increasing private sector activity • Experiments with formal value chain development • Land grabs (or long-term foreign land leases) • Land titling
Global shocks and a growing movement to decolonize African agriculture	2020–present	• Covid-19 pandemic supply chain disruptions and export difficulties • War in Ukraine: curtailed wheat imports; disruption of input imports	• Renewed discussion of agroecological alternatives in some cases • Business as usual in others	• To be seen

Source: Partially inspired by a chart of global food policy eras in Clapp and Moseley (2020).

(Carney 1993) and commodity crop production, such as cotton, cacao, coffee and tobacco (Bassett 2001; Moseley *et al.* 2015).

Malawi is a good example of a country that focused on increasing maize production in this era and The Gambia on rice production. While both countries increased their food crop production, there were challenges related to a narrow focus on one plant. In Malawi, the focus on high input maize varieties simplified the food system, reducing the historic importance of millet, sorghum and cassava (Bezner Kerr 2014). As farmers were increasingly encouraged to grow one crop, this led to labour bottlenecks in the agricultural calendar as planting, weeding and harvesting increasingly happened in one time period (Moseley 2000). Malawi's now maize-based food system also became increasingly reliant on inorganic fertilizer inputs, which would become a problem in subsequent eras when input subsidies were reduced (Chinsinga 2012; Jakobsen & Westengen 2022). In The Gambia, an aggressive plan was launched to double the country's rice output through the development of irrigated rice perimeters. By developing irrigation, farmers could produce over two seasons instead of one, with the thinking that one crop would be for household consumption and the other crop for sale. What planners neglected to consider was that women were the primary rice farmers and that this effectively doubled their work, leading to labour struggles within households (Carney 1993).

Modernization theory, the operative development philosophy of the time, also saw the commercialization of agriculture as a necessary first step in the industrialization process (Rostow 1960). This led to a rise in commodity exports, with the resulting tax revenue driving nascent industrialization on the continent. In the first decade of this era, African governments benefitted from a global commodity boom that allowed for investments in industrialization, such as government-run textile companies in many cases. A significant slowdown in the global economy in the 1970s, combined with rising energy prices, put substantial pressure on these state-run companies.

Neoliberal food security and trade (1980–2006)

As mentioned, by the mid-1970s the global economy began to slow, with a slackening demand for African commodities and rising energy prices. This led to growing indebtedness and the so-called Third World debt crisis by the late 1970s. This crisis would eventually lead to a set of reforms developed by international financial institutions (IFIs) known as structural adjustments policies (SAPs), which would be tied as conditions to multilateral lending moving forward (Berg *et al.* 1994). On the verge of defaulting on existing loans, many African governments would eventually take these new policy-based loans from the World

Bank and the International Monetary Fund (IMF). In exchange for such loans, African states took on SAP reforms that obligated them to: focus on a few export crops (for which they were deemed to have a comparative advantage); scale back government spending (including support for agricultural extension); remove tariffs and subsidies (including agricultural subsidies); devalue their currencies; and privatize industrialization efforts (which often led to deindustrialization).

For the next 25 years, agricultural spending was a low priority (and food self-sufficiency no longer a policy) for most African states as they focused on a few commodity crop exports and traded for any additional food that was needed. Neoliberalization also influenced land reform efforts launched in this era, such as the case of South Africa explored in Chapter 7. Here the World Bank's model of market-based land reform (also known as the willing seller/ willing buyer model) had a significant impact on the approach taken in that country. The Bank's approach advocated working through the market as more redistributive approaches were deemed to be too interventionist. Lastly, even in those countries that did not face a financial crisis and have to swallow the bitter pill of structural adjustment loans with conditions, such as Botswana discussed in Chapter 6, neoliberal thinking influenced that country's decision to abandon food self-sufficiency policies.

While I will explore the problematic impacts of structural adjustment on agriculture in Mali in Chapter 4, some scholars point to Ghana as an example of an African country where these policies were successful in the 1980s and 1990s (Loxley 1990). Although the country did experience some macroeconomic growth in this time period, the reality is that this growth was highly uneven, with southern areas of the country generally faring much better than northern areas (Konadu-Agyemang 2000). Furthermore, even though Ghana is one of the most prosperous countries in West Africa, its northern regions still face alarmingly high rates of food insecurity (Kansanga *et al.* 2022).

Neoproductionism and the new green revolution for Africa (2007–19)

The global food crisis of 2007–08 called into question a growing African reliance on food imports. With average food prices up 50 per cent, and some key commodities like rice up 100 per cent, there was a surge of food demonstrations, or food riots, across a range of African cities (Bush 2010). This social unrest caught the attention of African leaders and unexpectedly breathed life into a reboot of the Green Revolution known as the New Green Revolution for Africa. Starting in 2006, this initiative was similar to and different from the first Green Revolution. It was similar in its focus on the use of improved seeds, fertilizers and pesticides, but it would also be different in a number of ways.

These differences or innovations included a focus on African crops, attention to gender issues, a priority on public-private partnerships and an attempt to better integrate African smallholder farmers into global markets via a linear model known as a value chain (Annan 2007; Moseley 2017a; Gengenbach *et al.* 2018).

At the same time the New Green Revolution for Africa was kicking into gear, higher global food prices also spurred several foreign governments and companies to take out long-term leases on African land, sometimes known as land grabs (Moseley 2013a). Commercial farms were then established in these areas to produce for export. While these land leases (which often displaced African farmers) were portrayed by African governments as showcases for modern agricultural practices and sources of employment, the reality is that they were often a strategy of outside actors to produce food for their own populations or capture profits associated with increasing global food prices.

In Chapter 5 on Burkina Faso, I explore in some detail a New Green Revolution for Africa rice project. More broadly, the Alliance for a Green Revolution in Africa (AGRA) funded projects in some 18 African countries in this time frame (AGRA n.d.). Other donors were also inspired by this thinking and funded similar projects based on the value chain model, an approach that seeks to integrate smallholder farmers into broader markets by orchestrating connections between input suppliers, farmers, agro-processors and retailers. The results of these neoproductionist initiatives have not been promising with little to no positive impacts on household food security (Bassett & Munro 2022; Moseley & Ouedraogo 2022; Wise 2020). This approach has now further been called into question by the events of 2020–23 (namely the Covid-19 pandemic supply chain disruptions, the war in Ukraine, and related energy and input price fluctuations) where developing an energy-intensive production system seems ill advised.

Global shocks and a growing movement to decolonize African agriculture (2020–present)

In early 2020, Covid-19 infections began to spread around the world, impacting African food systems directly and indirectly. This led to lockdowns and disrupted trade, exposing the inadequacies of African agricultural development and food security strategies that had shaped African agri-food systems and were increasingly dependent on imported food, farming inputs and exports of commodity crops. Rates of hunger and malnutrition surged in many areas of the African continent as farmers lost export markets, food and input prices rose with declining currency values, and the urban poor struggled to buy food given income losses (Clapp & Moseley 2020). Ironically, some of those who weathered the pandemic best were subsistence-oriented farmers least impacted by development (Moseley & Battersby 2020).

Just as the Covid-19 pandemic was beginning to subside, the war in Ukraine broke out in early 2022. This was a problem for African food systems in two different ways. First, a number of African countries had become dependent on wheat imported from the Black Sea region, most notably populations in North Africa and the Horn (Moseley 2022a), and Ukrainian wheat exports declined precipitously in the 2022–23 period. Second, several African countries were also dependent on fertilizer imports from Russia and Belarus, which were disrupted by the war and rising insurance costs for shipping.

Given these recent disruptions, African policy-makers find themselves at a critical inflection point. Will they keep continuing down the path of developing energy intensive agriculture that is: vulnerable to global shocks; employs fewer people; undermines the natural resource base; is dependent on export markets; and has little impact on nutrition security at home? While the vision of an increasingly urban Africa is appealing, and the industrialization of agriculture is seen as the path to get there, policy-makers must ask hard questions about the modernist agenda they have been chasing since independence. Will these policies lead to full employment, result in nutrition security, and be resilient in the face of climate and global economic disruptions? While I wish the answers were different, government statistics suggest that the responses to all of these questions are a resounding no. Donors, international institutions and African policy-makers need to imagine a radically different path to future prosperity and well-being. Civil society organizations and some donors are now exploring an agroecological alternative that will be further examined in Chapter 3.

3

CONCEPTUALIZING CHANGE

New ideas are needed to rethink agricultural development and nutrition security in the African context, particularly a set of concepts and theories that will help analyse past problems and chart a new way forward. This chapter outlines five key frameworks that form the theoretical backbone of this book: intellectual de/colonization, political ecology, political agronomy, food security and agroecology. In the conclusion to the chapter, I discuss how these concepts work together in the African context to provide the understanding needed to foster change.

Intellectual colonization and decolonization

Ideas and theories shape how we see and act in the world. I did not always understand this. I remember tiring of theory by the end of my undergraduate education. At the time, I had this deep sense that the academy was totally removed from the real world and that I just wanted to go out and make a difference. So off I went into the Peace Corps. Early on in my Peace Corps tenure in southern Mali, I encountered what other scholars have referred to as forest islands, that is, patches of forest on an agricultural landscape that had seemingly been left undisturbed. I remember walking into such a forest island, looking up at the forest canopy and marvelling, then thinking that this must be a remnant of a previously forested landscape. While I did not process it as such at the time, my thinking in the 1980s was deeply influenced by certain conceptions of environmental change in the West African Sudano-Sahel, a collection of ideas I had learned in college (e.g. Harrison 1987) and certain outlets in the popular press such as *National Geographic* magazine (Ellis 1987; Moseley 2005). At the time, the dominant narrative suggested that the Sahara Desert was expanding and moving southward via a process of desertification (Swift 1996). This process was aided by human mismanagement of the environment that was driven in part by human population growth. In other words, what I saw as forest islands, and processed as remnants of a previously forested landscape, fit into my mental model of environmental change in the region. To put it even more bluntly, the theories that I had written off in my undergraduate days as disconnected from reality were actually shaping the way I processed and interpreted the world around me.

Fast forward a decade when I was in graduate school reading Fairhead and Leach's (1996) groundbreaking *Misreading the African Landscape: Society and Ecology in a Forest-Savanna Mosaic*. Working in Guinea, Fairhead and Leach (1996) showed that forest islands were actually created by people (and were not remnants of a previously forested landscape), something they established through careful ethnographic interviews with older community members and a time series analysis of aerial photography. Furthermore, they discussed the tenacity of regional environmental narratives dating back to the colonial period interpreting these forest islands as what was left of a forested landscape (the same narratives that had influenced my thinking). While this work was groundbreaking at the time, it was especially mind blowing for me as I had lived it. In fact, the desertification narrative had shaped the way I was trained as a Peace Corps volunteer (Moseley & Laris 2008). We were part of what was called the African Food Systems Initiative (AFSI) and, working in teams of gardeners, foresters and water volunteers, our task was essentially to roll back the desert and promote, as then Peace Corps Director Ruppe (1985) put it, "the greening of Africa". I would subsequently come to understand that I had been an actor in a narrative that had existed since the French colonial times in the region. To this end, Fairhead and Leach (1995) cite the work of French colonial era botanist Auguste Chevalier (1909) who wrote about the fire-setting practices of local people that had turned the upland forest of Guinea into more of a savanna landscape. They also cite the work of Adam (1948) who discussed how the Mandinka (a local ethnic group) "were a savanna people who had migrated southward into the forest zone, and created savanna there" (Fairhead & Leach 1995: 1024).

Some of the ideas and models that are central to our disciplines date back to the colonial period and may be based on shaky science, racial prejudice, a belief that European regional experiences could be universalized, or the prerogatives of colonial empires (Leach & Mearns 1996). While these ideas may be wrong or inaccurate, they have been perpetuated over time and received little scrutiny. Although this is a problem in all disciplines, including my own discipline of geography (Maharaj & Ramutsindela 2021), I would argue that it is a particular problem in some of the applied sciences such as forestry, agronomy, water resources management and public health, because it is these sciences that were often closest to the colonial enterprise. As Leach and Mearns (1996) have argued, if the science is flawed, then it very challenging to solve problems and move forward.

Figure 3.1 Removing a statue of Cecil Rhodes at the University of Cape Town campus in South Africa
Source: Desmond Bowles, Wikimedia Commons, https://creativecommons.org/licenses/by-sa/2.0.

As such, a shift in agricultural and food system practice will require a shift in the ideas and world views that guide practice, or an intellectual decolonization of the agricultural sciences. Intellectual decolonization is different than political decolonization, the liberation process that unfolded in Africa in the post-Second World War period. Rather, intellectual decolonization[1] is about recognizing the assumptions and Eurocentric ideas that are pivotal to many disciplines. This is not a new realization, for example Paulo Freire was writing about the decolonization of the mind and critical pedagogy in the 1960s (Freire 1982), as was Ngugi Wa Thiong'o in the 1990s (1998), particularly with respect to the importance of writing in African languages (as language shapes thinking). Nonetheless, applying this approach to the agricultural sciences is more recent (Sumberg 2017; Moseley 2021a).

More broadly, intellectual decolonization is an ongoing process in the academy today. The image in Figure 3.1 is from the "Rhodes Must Fall" demonstrations in South Africa. This was a student protest movement that started

---

1. A closely related term is decoloniality. "While decolonization generally refers to the dismantling of colonial systems, structures or institutions – including universities and research institutes – decoloniality is about a way of "re-learning" knowledge that has been pushed aside, forgotten or discredited by forces born of modernity, settler colonialism and race-based capitalism. Unlike decolonization, decoloniality is a method and a paradigm for recreating and perhaps even repairing that which depends on context, historical conditions and geography" (based on a personal conversation with Sten Hagberg, Professor of Anthropology, University of Uppsala, 24 July 2023).

on the University of Cape Town (UCT) campus in 2015 to remove a statue of Cecil Rhodes. Rhodes was a mining magnate, supporter of British colonialism and deeply problematic figure whose legacy had been celebrated by white settler colonialism in southern Africa. The Rhodes Must Fall movement caught the attention of South Africa and the world, leading to a growing recognition that higher education in South Africa was deeply colonial (Ndlovu-Gatsheni 2018).

Political ecology

Political ecology has been broadly defined as the political economy of human–environment interactions (Blaikie 1985; Robbins 2004). There are three aspects of the political ecology perspective that are most relevant to this book, including the importance of: (1) thinking across different scales of policy and power; (2) the role of discourse; and (3) socially constructed gender roles.

First, political ecologists understand that the way humans interact with the landscape is often mitigated and shaped by broader political economy, power and politics. A farmer does not just decide to grow cotton using a certain package of inputs (improved seeds, herbicides, insecticides and inorganic fertilizers), to prepare the land in a particular manner, or to sell his or her crops to certain markets. Rather, this farmer's action, or agency, is situated in a particular context and these decisions are shaped by a variety of policies and interests at the local, national, regional and international levels. As such, the cotton farmer in Mali is influenced by policies and programmes of the national government, regional trading relationships, and the policies of international institutions. Furthermore, this farmer's experience of environmental variability and change, which we often think of as an exogenous factor, is also shaped by broader political economic and historical factors. Our farmer's experience of hunger in a drought year may be shaped by the fact his family no longer stores grain from preceding surplus years (which would buffer such impacts) but must face the full brunt of their precarity given a farming system that is focused on maximum production rather than risk management (Watts 1983).

Second, the discourses and theories that shape our understanding of, and actions in, the world (discussed above under intellectual decolonization) have been shaped by powerful colonial and postcolonial interests (Peet & Watts 2002a). In other words, powerful entities not only shape the realities on the ground in a material sense, but they have discursive power or the ability to shape the narratives that influence thinking in the development community (or other communities of practice). For example, while the World Bank is widely known for its development-oriented lending operations, it also has a huge research division (where I happened to once work in the Environment Department), which churns out publications like its annual development report. Whether or not these publications influence World Bank lending operations, these reports have significant discursive power, shaping the way the development community thinks about certain issues. To wit, the World Bank's (1992) annual development report on environment and development helped give rise to the idea that poverty leads to environmental degradation, a line of thinking that subtly blamed the world's poor for environment destruction (Gray & Moseley 2005). Of course, the idea that powerful interests may shape knowledge and discourse is not new (see Foucault 1971, 1980), and that certain understandings may become hegemonic or widespread (Gramsci 1971), but this thinking became more integrated into post structural political ecology frameworks in the 1990s courtesy of scholars like Peet and Watts (2002b).

The other aspect of discourse and post structural political ecology that is important to note is the role of agency and social movements. Older political ecology thinking, sometimes referred to as structural political ecology (Moseley *et al.* 2013), was subtly defeatist about local people's ability to change matters in stressing broader scale forces or political economy. In emphasizing the role that discourses play in shaping our understanding of reality, a new space is created for social movements to disrupt and reshape these narratives or to create counter narratives. As such, for example, the food sovereignty movement has pushed back against conventional understandings of agricultural development, giving rise to new understandings of agricultural change and food security (Nyéléni Declaration 2007).

Third, political ecologists highlight the role of socially constructed difference (in terms of gender, class, race, ethnicity, etc.) in shaping how people interact the environment. More specifically, gender is an important construct in many African countries, where women are more likely, for example, to collect firewood or fetch water in rural areas, making them more attuned to the degradation of these resources. This attention to gender has given rise to a subfield of political ecology known as feminist political ecology (Rocheleau *et al.* 2013). The highly gendered nature of African farming and food systems means that women often: perform certain farming tasks (Schroeder 1999); farm particular crops (Carney 1993); engage in more foraging (Morgan & Moseley 2020); are active in informal food markets (Clark 1994); raise certain animals (Hovorka 2006); and may have limited access to land and inputs without going through male relatives (Nyantakyi-Frimpong 2017; Moseley & Ouedraogo 2022).

Political agronomy

The concept of decolonization, and more broadly knowledge politics, is central to an interdisciplinary conversation regarding political agronomy. The main idea is that what we call agronomy or crop science is not apolitical, but inflected with politics. As Sumberg *et al.* (2012) and Ross (2014) have suggested, development-oriented agronomy, originally tropical agronomy, has its roots in Europe and was driven by the need for colonial powers to capture and modify tropical crops, soils and farming practices as fuel for European economic expansion. In the postcolonial period, Sumberg *et al.* (2012) further argue that development-oriented agronomy has been influenced by the neoliberal project. Understanding knowledge politics is also central to political agronomy (Andersson & Sumberg 2016). As Vanloquerin and Baret (2009) note, an analysis of knowledge politics in agronomy helps explain why particular technologies or development pathways are privileged over others.

In terms of specific examples, agronomy has long held a relatively privileged position in the French academy and played an outsized role in the French colonial enterprise. Since the early nineteenth century, the French higher education system has operated with two tiers of schools, the *grands écoles* and regular universities. The former are considered to be more elite institutions charged with training leaders for the civil service and business communities (Bret 2002). Along with public administration, engineering, forestry, business and other fields, agronomy has its own *grand école*, the *Institut National Agronomique* (Graves 1965). The *grands écoles* produced many of the civil servants who oversaw the French colonial apparatus, including agronomists. The goals of French agronomists were pretty consistent throughout the colonial period, namely to produce agronomic research that would help the empire extract more wealth from its colonies. This came in the form of developing better seeds, combatting diseases and improving yields for certain crops such as palm oil, coffee, peanuts and cotton (LaFargeas 2019). Much of this research was spread across nine crop or commodity chains (*filière* in French), related tropical research institutes that today have been reorganized under the French Agricultural Research Centre for International Development (CIRAD), based in Montpellier, France. As such, while CIRAD's contemporary work is much more interdisciplinary and progressive (as I discovered when I spent a semester at CIRAD as a research fellow), it is a direct descendent of the colonial agronomy enterprise (Bonneuil & Kleiche 1993). This is not to suggest that the British or American agricultural development apparatus is any better or less culpable, as agronomy in each of these countries also has its own story.

A six-dimensional understanding of food security

While agronomists have long focused on increased food production (or availability) as the best way to address hunger, there is a growing consensus in scholarly and policy communities that food security has not one but six dimensions that need to be considered: availability, access, utilization, stability, sustainability and agency (HLPE 2020; Clapp *et al.* 2022). This broadened understanding of food security has been incrementally developed over time, with an increasing recognition that all six dimensions need to be addressed via policy and programmes if food security and nutrition is to improve. Herewith a slightly more in-depth discussion of the six dimensions of food security that were briefly introduced in Chapter 1.

Availability

This refers to the amount of food available on the market or how much food is available in a certain jurisdiction. It is a combination of food that is produced locally as well as net imports (imports minus exports). Availability may be conceptualized at a variety of scales (household, neighbourhood, city, province, country). This has been the traditional focus of much mainstream food security work in the postcolonial period (if countries produce more food then they will solve the hunger problem), not to mention early famine early warning systems, such as the food balance sheet approach that compared food supplies to the caloric needs of a population (Moseley & Logan 2005). A narrow focus on availability ignores the other dimensions of food security. Critics have raised concerns about the Green Revolution or community gardens as overly focused on food production that is linked to local food availability. Furthermore, in an effort to address hunger by producing more food, countries have often created food systems that are more vulnerable and less resilient (Moseley 2022b). For example, crop monocultures are often more susceptible to drought than polycultures (Altieri *et al.* 2015).

Access

This refers to the resources that people have to access food. There may be plenty of food on the market (availability), but this will not address malnutrition if the food is too expensive and/or people do not have the funds to purchase the food. Amartya Sen (1982) conceptualized this in terms of his concept of "entitlements". People may have legal entitlements to food by producing it themselves, purchasing it at the market, via reciprocal relationships with extended family and neighbours, or with government support. According to Sen, most famines/hunger are not about lack of food availability but entitlement failure (lack of access). For example, the famine in the Sahel in the mid-1970s had as much more to do with entitlement failure than production declines (Franke & Chasin 1980). Increasing awareness of food access as a problem has also led to a growing literature on food deserts in urban areas, a situation where poorer neighbourhoods have fewer food retail outlets. While more food outlets in urban areas (often driven by the supermarketization phenomenon in the Global South (Reardon & Hopkins 2006)) might appear to be a solution, Peyton *et al.* (2015) have shown that a growth in formal supermarket chains in South African cities has sometimes crowded out small informal shops that provide vital services to the poor, such as the sale of food on credit or in small quantities.

Utilization

Utilization refers to the proper energy, water and sanitation infrastructure needed to clean and prepare food. Cooking knowledge is also important. The Food and Agriculture Organization (FAO) defines food utilization as "the proper biological use of food, requiring a diet providing sufficient energy and essential nutrients, potable water, and adequate sanitation. Effective food utilization depends in large measure on knowledge within the household of food storage and processing techniques, basic principles of nutrition and proper childcare" (FAO 2006: 1). This dimension of food security is arguably the most under-recognized, but it can have impacts on people's diets and consumption patterns in surprising ways. For example, rising cooking fuel costs in South Africa have led to increased purchases of less nutritious, pre-prepared foods (Moseley & Battersby 2020). Limited access to proper sanitation, clean water and cooking facilities may also be hastening the nutrition transition in the continent's urban areas, a situation where people's diets are moving from traditional foods to more sugar, meat and processed food, leading to a rise in non-communicable diseases such diabetes and hypertension (Nnyepi *et al.* 2015).

Stability

Stability pertains to the regularity of food supplies. While global food supplies and prices were fairly stable in the 1980s and 1990s, they have become more variable from the 2000s, with price spikes in 2007–08, 2011–12 and 2021–22. The 2007–08 food price spike was commonly referred to as the global food crisis where average food prices went up 50 per cent and the price for rice increased by 100 per cent. This led to a lot of social unrest in some African cities (Moseley *et al.* 2010). Furthermore, many supply chains broke down during the Covid-19 pandemic, leading to bare supermarket shelves in some cases (Clapp & Moseley 2020).

Sustainability

Sustainability in the context of food security refers to the long-term ability of food systems to provide food security and nutrition in such a way that does not compromise the economic, social, and environmental bases that generate food security and nutrition for future generations (HLPE 2020; Clapp *et al.* 2022). Farming approaches that undermine the environment work against food security in the long run and across generations. For example, the Office du Niger irrigation scheme, a major rice growing area in central Mali, has experienced salinization problems related to improper drainage that damaged soils and undermined production in the long term (Bertrand *et al.* 1993).

Agency

Agency refers to the capacity of individuals or groups to make their own decisions about what foods they eat, what foods they produce, how that food is processed and distributed within food systems, and their ability to engage in processes that shape food system policies and governance (HLPE 2020; Clapp *et al.* 2022). Agency is also central to food sovereignty and certain conceptions of agroecology. While food sovereignty is often associated with social movements in Latin America, African food and peasant movements have also been at the forefront of defining and shaping this concept. In February 2007, more than 500 representatives from organizations representing peasants, family farmers, fisher folk, indigenous peoples, landless peoples, rural workers, migrants, pastoralists, forest communities, women, youth, consumers and environmental and urban movements gathered in the village of Nyéléni,[2] near Sélingué in southern Mali. From their deliberations emerged the Nyéléni Declaration, a seminal statement on the principles of food sovereignty. They defined food sovereignty as "the right of peoples to healthy and culturally appropriate food produced through ecologically sound and sustainable

---

2. Other than the name of the village where the declaration was signed, Nyéléni also means "first daughter" in Bambara or Bamanankan. According to oral history, Nyéléni "was the only child of a Malian peasant couple at a time when having only one child, and a daughter at that, was considered shameful. Despite these attitudes, Nyéléni became a highly regarded farmer who supported her own family and many others through her hard work and ingenuity in production and processing of food. She is credited with the development of a local grain called "fonio" that, several hundred years later, is still an important food crop. The symbolic presence of this iconic figure at the Forum that was given her name was particularly important for the many women food producers there. As farmers, foragers, herders, processors and cooks, women in Africa as elsewhere have a central place that was recognized and honoured at Nyéléni 2007" (Nyéléni Declaration 2007: 13).

methods, and their right to define their own food and agriculture system" (Nyéléni Declaration 2007). By recognizing agency as a dimension of food security, I would argue that we link it to food sovereignty in productive ways.

Agroecology as a new paradigm for moving forward

The way African farmers produce food and connect to markets is being reimagined along the lines of agroecology (Nyantakyi-Frimpong & Bezner Kerr 2015; Bezner Kerr *et al.* 2019) and food sovereignty (Nyéléni Declaration 2007), especially as we come to understand the deeply colonial roots of the conventional productionist paradigm (Sumberg 2017; Moseley 2021a). The importance of urban governance, gender empowerment and a robust civil society to food systems is also becoming increasingly clear (Battersby & Watson 2018).

As a science, agroecologists treat the farm field as an ecosystem, studying the ecological interactions between different crops, crops and insects, crops and animals,[3] as well as crop, climate, soil interactions (Carroll *et al.* 1990). By understanding and leveraging synergies within these systems, the farmer may improve production, limit labour requirements, minimize insect damage and maintain soil fertility. At the end of the day, by working with nature – rather than against it – one produces more with less expensive inputs, making this approach accessible to the poorest of the poor, the people who are most prone to food insecurity. This approach is also more sustainable than conventional measures and its embrace of indigenous knowledge enhances local agency.

Two practices that are central to agroecology are intercropping and agroforestry. One example of intercropping (aka polycropping) in African farming systems is the frequent combination of sorghum and cowpeas. Cowpeas are a legume and fix nitrogen that is used by sorghum. More diverse polycultures also tend to suffer less from insect predation than monocultures. Another common practice in agroecology is agroforestry or the mixing of trees and crops in the same field. Parkland agroforestry is a very common practice in many African farming systems, wherein trees of ecological or economic value are interspersed throughout a farm field. One example of a commonly used agroforestry species in the West African Sudano-Sahel is *Acacia albida* (see Figure 3.2). *Acacia albida* is a leguminous tree that fixes nitrogen in the soil to the benefit of crops planted in the understory. It also loses its leaves during the rainy season so that it does not compete with crops for sunlight. During the dry season the tree leafs out, protecting the soil from solar radiation and the decomposition of organic matter.

---

3. A classic example of this is crop–livestock integration wherein animal manure is used to fertilize crop fields and crop residues or byproducts are used to feed animals (Moseley 2022c).

Figure 3.2 Example of an agroforestry system in Burkina Faso: *Acacia albida* trees with emerging crop early in the rainy season
Source: Photo by author.

It also cuts down on windborne soil erosion during the dry season, a time when dry Harmattan winds are prone to creating dust and sandstorms. Last but not least, the *Acacia albida* produces a large pod that may be used as animal fodder.[4]

What these two practices (intercropping and agroforestry) also demonstrate is the more decolonial nature of agroecology, an approach that straddles two knowledge systems: traditional or experiential knowledge; and formal knowledge or science. Both of these practices were developed by African farmers through trial and error over time (suggesting that farmers are also scientists). Agroecology validates these practices, helps us understand why they work in terms of ecology and also allows for further experimentation with them in the formal science community.

Four questions sometimes arise with respect to agroecology. These include: (1) may it be "scaled up" or mass replicated; (2) does it imply going back to indigenous or traditional crop varieties; (3) are its labour requirements higher; and

---

4. For another example of agroforestry, see this short video I filmed in July 2019 discussing agroforestry examples in a farm field in southwestern Burkina Faso: https://youtu.be/qGjwE9PbviE?si=b0hgIacZ9tOoUNfs.

(4) will it allow African farmers to shift away from livelihoods that are dependent on the export of commodity crops, such as cotton and coffee? With respect to the first question, "scaling up" is a term that comes out of the business community. It often implies piloting an idea or programme, and then mass replicating it regardless of local context and knowledge (Moseley 2017b). Agroecology involves working with local conditions and knowledge systems, rather than a one size fits all approach. However, with regards to the second question, agroecology does not suggest simply going back to traditional crop varieties and practices. Rather, it is a blend of old and new practices that share an emphasis on leveraging positive ecological interactions that lead to improvements in the sustainable production of healthy food. In terms of the third question, this is an important concern because many African farming systems are labour constrained. While the jury is still out on this question, a study by Laske (2021) in Senegal found that agroecological farming did not require any more labour than conventional approaches in that country. Lastly, with regards to the final question, farmers do need to sell some crops to make money to cover household expenses, such as school fees and healthcare. What is called for is a rebalancing, a healthy mix of food and cash crop production, as well as production for local and external markets. For too long, the emphasis has been on external markets and this has created a food system that is vulnerable to shocks. It is also true that many commodity crops are amenable to agroecological practices, such as shade grown coffee and cacao systems (Toledo & Moguel 2012).

While I initially learned about agroecology as a science, agroecology is also a social movement that has close links to food sovereignty. This is important for two reasons. First, given the insights of political agronomy, it is increasingly understood that the agricultural sciences do not exist in a vacuum but are inflected with politics and situated within political economy. As such, while it might be rational to pursue the science of agroecology in order to develop healthier, more equitable and sustainable food systems, social movements are needed to reshape political economy in a way that will allow this new science to more fully emerge. Second, we just cannot graft agroecology onto the existing structure of farming systems in many cases, but the social organization of agriculture also has to be considered. While this may be less of a challenge in African countries where there is a lively smallholder farming sector, a social reorganization will be more challenging in places where land distribution is more uneven such as Zimbabwe and South Africa.

While agroecology has been at the margins for several decades, it has more recently gained traction as a more mainstream discourse in Europe and the United Nations system (Loconto & Fouillieux 2019; HLPE 2019). This growing acceptance means that EU funding is available in some cases for experimentation with agroecology. There are also examples of African civil society organizations in support of agroecology, such as the Alliance for Food Sovereignty in Africa (AFSA n.d.). Last but not least, in addition to ideas, funding from the donor community and support from civil society, African political leaders are needed to foster the emergence of a new paradigm. There are historical examples of this, such as Thomas Sankara, the Marxist political leader of Burkina Faso in the 1980s who heavily promoted self-sufficiency and agroforestry (Jacobs 2013; Leshoele 2019). More recently, the Senegalese Minister of Agriculture (now ambassador to Italy), Papa Abdoulaye Seck, embraced agroecology during his tenure as minister and has pushed it as ambassador in his deliberations with the UN's Rome-based agencies, the FAO, WFP and IFAD. What is remarkable about Ambassador Seck is that he was trained as an agronomist and long supported the neoproductionist, New Green Revolution for Africa approach. He then had an epiphany based on several years of observation, shifting to become supportive of agroecology (Boillat *et al.* 2022).

Conclusion

These five frameworks, intellectual de/colonization, political ecology, political agronomy, a six-dimensional understanding of food security and agroecology, can be seen as working together in the African context to conceptualize, understand and foster change (Figure 3.3). The first three frameworks (intellectual de/colonization, political ecology, political agronomy) help provide the concepts and ideas needed to better understand why African agricultural development

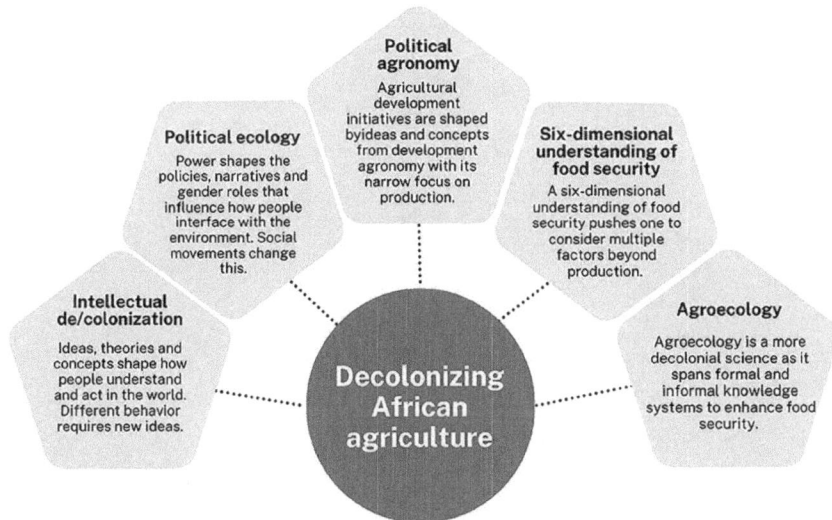

Political agronomy
Agricultural development initiatives are shaped by ideas and concepts from development agronomy with its narrow focus on production.

Political ecology
Power shapes the policies, narratives and gender roles that influence how people interface with the environment. Social movements change this.

Six-dimensional understanding of food security
A six-dimensional understanding of food security pushes one to consider multiple factors beyond production.

Intellectual de/colonization
Ideas, theories and concepts shape how people understand and act in the world. Different behavior requires new ideas.

Decolonizing African agriculture

Agroecology
Agroecology is a more decolonial science as it spans formal and informal knowledge systems to enhance food security.

Figure 3.3 Conceptual map for decolonizing African agriculture
Source: Figure conceived by author, with artwork by Kellen Chenoweth, Macalester College.

efforts have failed to provide better nutrition and food security, whereas as the last two (a six dimensional understanding of food security and agroecology) point the way forward for a more decolonized African agriculture and a healthier and more sustainable future.

The next section of the book dives into four country level studies, in Mali, Burkina Faso, Botswana and South Africa, exploring how conventional agricultural development has failed to deliver on the promise of better food security and nutrition.

PART II

COUNTRY STUDIES OF FAILED AGRICULTURAL DEVELOPMENT IN THE COLONIAL AND POSTCOLONIAL PERIODS

4

MALI: FROM WEST AFRICAN BREAD-BASKET TO MALNUTRITION AND THE COTTON COMMODITY TRAP

In 1998, Mali was the leading cotton producer in all of Africa, culminating in ten years of steadily increasing production (Badiane *et al.* 2002).[1] Not only that, Mali had developed a reputation for producing some of the highest quality cotton in the world, known for being hand-picked with long, strong fibres. In many ways this was a triumph of the vertically integrated Malian Cotton Company (CMDT) system that facilitated and coordinated the development and distribution of high-quality cotton seed, farm extension services, the sale of inputs to farmers on credit, and the collection and ginning of the country's cotton crop. In that year, the CMDT parastatal, jointly owned by the Malian and French governments, generated some 40 per cent of the revenues for the Malian state, leading to wide spread recognition among civil servants that cotton paid the bills (even leading some to refer to the crop as Mali's "white gold") (Moseley 2001). At the centre of this cotton boom was the country's southernmost region, Sikasso. This region was historically known as the country's bread basket with plentiful rainfall, abundant land, verdant gardens and hard-working farmers. Yet, curiously, in this same period, nutrition researchers were discovering that this wealthiest and most productive of Mali's agricultural areas also suffered from some of the highest rates of child malnutrition in the country, a problem that would come to be known as the "Sikasso paradox" (Tefft *et al.* 2000; Dury & Bocoum 2012).

This juxtaposition of a highly productive system for commodity crop production and malnutrition lies at the centre of this story about agricultural development in Mali. This chapter explores the role of cotton in Mali's agricultural development, from the precolonial period until the present, including broader changes in the country's food and farming systems. After presenting

---

1. For context, Mali was also the leading cotton producer in Africa in 2021–22 after several years of below average production. Public subsidies for inputs (seed, fertilizer and pesticides) played a key role in increasing the area devoted to cotton and resurgent production (Africanews 2022).

some contextual information on the country's geography and history, I review the farming system changes that occurred in different periods as well as the specific environmental and social problems related to these shifts. This analysis is informed by the theoretical frameworks presented in the previous chapter, namely intellectual colonization, political ecology, political agronomy, food security and agroecology.

Understanding Mali

Some readers may only have a limited familiarity with Mali. This landlocked nation of some 22 million people is located in the centre of West Africa. Covering 1,240,190 square kilometres, Mali is the eighth largest African country in terms of surface area. It has three major climate and vegetation zones: desert in the north; semi-arid or Sahel grasslands in the centre; and more humid or savanna grasslands in the south. The majority of the country's population resides in the lower third of the country, below the Sahara Desert, and along major river courses. Average rainfall in the desert area ranges from 0 to 350 mm, in the Sahel zone from 350 to 600 mm and in the savanna region from 600 to 1,200 mm (see Figure 4.1). The southern part of the country (Sahel and savanna regions) has two distinct seasons, a rainy season from roughly June to October and a dry season from November to May. The dry season is divided into a cold season (December–February) and a hot season (March–May). Dusty harmattan winds from the North often spread across the country during the dry season (Moseley 2007a; Baker *et al.* 2023).

Mali's majority rural population is composed of herders in the drier areas, farmers in the grasslands and fisherfolk scattered along various rivers, with the Niger River being the most important. The major food crops are sorghum, millet, maize, rice, peanuts and cowpeas. Cotton is the major cash crop. Due to more rapid urbanization in recent years, the country's urban centres now comprise some 40 per cent of the population, with the largest city being the capital, Bamako. The country is ethnically diverse, comprising some ten ethnic communities of which the Mande groups (Bambara, Malinke, Soninka) are the most numerous (50 per cent). Major exports include gold, cotton and livestock (Moseley 2007a).

While much of the contemporary news about Mali skews negative, often highlighting drought, conflict and hunger, it is important to know that the country has a long and storied history, being home to several great medieval African kingdoms, including the Empires of Ghana, Mali and Songhai. The Mali Empire (and namesake for contemporary Mali) existed from roughly 1230 to 1545, and stretched from Senegal's Atlantic coast to the Sahelian town of Gao

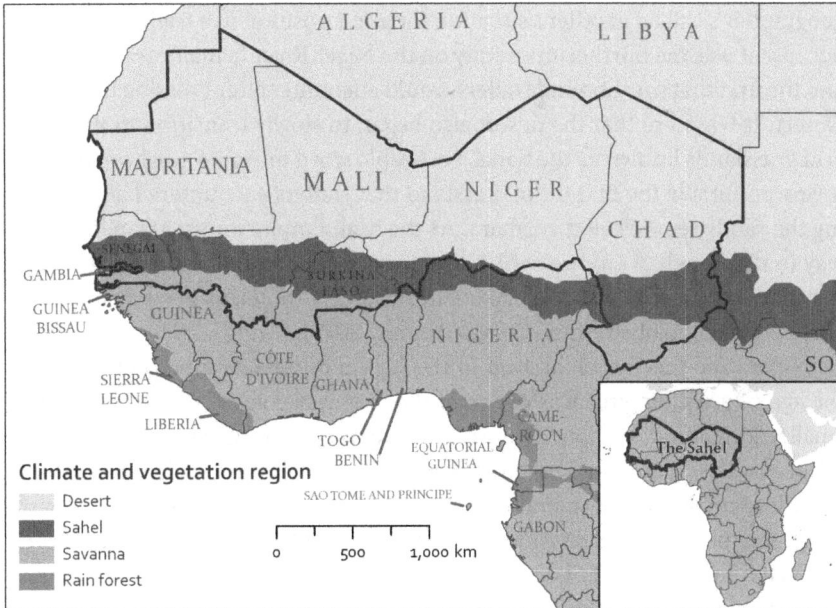

Figure 4.1 West Africa's vegetation and climate zones
Cartography by Kellen Chenoweth, Macalester College.
Sources: Esri Africa 2018; Humanitarian Data Exchange 2018; ESRI Africa 2018; World Bank 2023f; Africa Albers Equal-area Conic Projection

at its height.[2] The source of the empire's wealth was control of gold mines to the south, salt mines in the north, and trade across the Sahara Desert. The commercial and intellectual centre of the Malian Empire was Timbuktu,[3] located at the apex of the Niger River, a waterway which originates in the highlands of Guinea, makes a huge arc into the Sahara Desert, before flowing south again through present day Nigeria and to its outlet in the Gulf of Guinea. Timbuktu had what

2. Mali's most notorious medieval king was Mansa Musa or King Musa (r.1307–32) who some assert was the wealthiest man to have ever lived. He famously travelled to Mecca on a Haj pilgrimage, spent lavishly in Cairo, and established one of the world's oldest universities in Timbuktu (known as the Sankoré Madrassa or University) stocked with books and scholars who returned with him from Mecca. Due to Mansa Musa's generous patronage, the Sankoré Madrassa held some 250,000–700,000 manuscripts, making it the largest archive in Africa since the Great Library of Alexandria (burned by Julius Caesar in 48 BC) (Conrad 2009). In more recent times, Malian stewards of this library have risked life and limb to hide and protect these manuscripts from fanatical Islamists who sacked the Madrassa, seeking to destroy its books (Hammer 2016).

3. The administrative capital of the Mali Empire was the city of Niani, located in the highlands of contemporary Guinea, near the border with present day Mali. Niani was surrounded by the Bure goldfields, an important source of wealth for the empire.

geographer's call an excellent situation (or good position in a trading network), because it was the northernmost city on the Niger River (a major transit artery) and the first that southbound traders would encounter after crossing the Sahara Desert. It was here that the desert also began to slowly transition to the semi-arid grasslands known as the Sahel, an Arabic word meaning seashore because it was essentially the first bit of grassland that traders encountered after crossing the sandy sea we call the Sahara. As the Mali Empire's power waned, it gave way to the Songhay Empire, which took over much of the Mali Empire's territory and expanded it further, making it the largest (in terms of surface area) of Africa's precolonial empires (Canós-Donnay 2019).

By the time the French arrived in the region in the late nineteenth century, the age of great empires was over and the area was controlled by a number of smaller kingdoms, such as Samori Toure (discussed in Chapter 2) and Tieba Traore in the South, and Sekou Ahmadou in the central Macina area. While these leaders put up considerable resistance to colonial armies, the French were adept at exploiting local divisions, and the area eventually succumbed to French colonial rule in the 1890s, becoming known as the French Soudan, and administered as part of French West Africa (which included the contemporary countries of Senegal, Mauritania, Guinea, Mali, Niger, Côte d'Ivoire, Burkina Faso and Benin). While Senegal, Côte d'Ivoire and Algeria were more prized colonial possessions of the French in the region, Mali was valued for its key location on the Niger River (which spans the entire country), productive agricultural capacity and hard-working population.[4]

Mali achieved independence in 1960, the result of the pan Africanist movement and a French state greatly weakened by the Second World War. After a brief federation with Senegal, which some believe the French actively worked to undermine, Mali broke off and increasingly drifted towards the Soviet bloc under the leadership of its first president, Modibo Keita. While Keita tried to foster local industries (consistent with the modernization theories of the time), his undemocratic tendencies increasingly put him at odds with the population. Keita succumbed to a military coup in 1968, with General Moussa Traoré ascending to the presidency and ruling until 1991. Traoré was a somewhat lacklustre leader who managed to

---

4. Unlike more valuable colonies, the French were not very numerous in Mali. In fact, an elderly Malian friend once told me that he never saw the French growing up as a child in the late colonial period. The French had an administrative approach known as direct rule wherein French colonial officials held nearly all of the top administrative positions, managing countries in a more or less militaristic and top-down fashion. This contrasts with the British approach of indirect rule where they often ruled through African intermediaries (Moseley & Otiso 2022). Nonetheless, in Mali most people would have interacted with the government at the canton level (the lowest administrative unit similar to a district or county), and here the French regime would have been represented by local Malian leaders given the limited number of French administrators.

keep the peace, yet presided over growing corruption and ushered in the neolib-eral reforms of the 1980s (shrinking the size of the government) when the country had to accept World Bank and IMF loans or risk defaulting on its debt. Popular uprisings in 1991 led to a military coup, a year-long interim government, and the popular election of Alpha Oumar Konaré in 1992.[5] Konaré was arguably Mali's most successful president, being democratically elected and serving two five-year terms until 2002. He is known for his restoration of democracy in the country, rela-tively successful management of long simmering ethnic troubles in the north with the Tuareg ethnic group, and decentralization (a programme to push out control and elections to lower levels of government). It is also notable that he voluntarily left power at the end of his second term (as Mali has a two-term limit). His biggest shortcoming was that he never managed to address corruption within the govern-ment at a time when foreign assistance was pouring into the country.

In 2002, Alpha Konaré was succeeded by Amadou Toumani Touré (often known by his initials ATT). While democratically elected, ATT had a military background and ruled until March 2012 when he was ousted in a coup. The coup was driven by a military unhappy with the management of the conflict in the North and popular discontent with corruption. The 2012 coup was a major turning point for the country, not only ending 20 years of democratic rule (which some view more sceptically in retrospect) but the launch of a new phase of pro-tracted instability and weak governance, which included the temporary loss of the northern part of the country to jihadist rebels and a ten-year stay by French and United Nations Peacekeepers (2013–22) who struggled to contain jihadist insurgencies and violence across the country. During this time period, military coup leader Amadou Sango led the country for a year in 2012, allowing for the democratic election of Ibrahim Boubacar Keïta (known as IBK), who ruled until 2020 when there was yet another coup (again driven by military discontent with the conflict in the north). Colonel Assimi Goïta has been interim president of the country since that time. Goïta's most controversial moves were to ask the French military to leave the country and to hire the Russian Wagner Group for protection (Gazeley 2022).

Farming, food and cotton in the precolonial period

While most of Mali's agriculture was subsistence-oriented in the precolonial period, it was productive enough to support fairly large cities and towns in the

---

5. Alpha Konaré was an archaeologist and Professor of History and Geography who had a history of political activism and involvement, stressing – among other things – the importance of a free press.

age of the medieval kingdoms discussed above. According to the accounts of Arab travellers, Niani (the capital city of the ancient Mali Empire) had about 100,000 people in the fourteenth century, Gao (the capital of the Songhay Empire) had a population of 76,000–100,000 in the sixteenth century, and Timbuktu numbered around 80,000–140,000 people in this period as well (McIntosh 2005). Furthermore, Arab writers report an abundance of food and provisions in Mali's villages. Their food production would not only have supported the aforementioned urban populations but a standing army of 100,000 men. The king or mansa was also known to throw large public feasts. This food and grain would have been supplied to the authorities via a system of tribute (Canós-Donnay 2019). There are at least three aspects of Malian agriculture in the precolonial period that would have made it highly productive.

First, according to Widgren (2012), many Malian dryland farmers employed a strategy known as ring cultivation (or in-field, out-field system) wherein significant amounts of organic waste and manure were applied to fields close to the village, with declining intensity as one moved away from the community. This led to the development of what is known as anthropogenic soils or highly productive fields that were the result of active human management. Multiple crops would have been grown in association, leading to high levels of productivity when all crops are considered, and on much smaller fields than today given the labour demands of clearing and tilling. Most farming households would have been patriarchal, extended families (grandparents, sons, wives and children) as this allowed the male head of household to harness the labour needed to run a successful farm (Lewis 1979). Lastly, grain storage was an integral part of the system given yearly fluctuations in rainfall, with most households storing a two to three-year supply of grain (Widgren 2017).

Second, African rice, which is different than Asian rice that predominates around the world today, was originally domesticated in the inner Niger delta (Carney 2002). This is a massive inland delta, created by a naturally low-lying area, found in central Mali to the southwest of Timbuktu that seasonally floods when river levels rise during the course of the rainy season. The delta provided the perfect natural conditions for floodplain rice production. The annual influx of water led the river to spill over its banks, flooding hectare upon hectare of rice fields. Furthermore, the annual deposition of silts continuously replenished the fertility of the soil. This very same delta also provided rich pastures for livestock when flood waters receded and a robust fishery. Effectively managing the natural resources in this area took some coordination. In the nineteenth century, between 1820 and 1862, we see the emergence of the Fulani Dina or kingdom in the inner Niger delta under the leadership of Sekou Amadou (mentioned previously). While primarily serving the interests of Fulani herders, one of the chief functions of the Dina was to coordinate the overlapping use of the land by farmers, herders and fisherfolk in such a way that conflicts were minimized and production was enhanced (Moorehead 1991; Moseley *et al.* 2002). Such coordination would likely have led to synergies that we see from crop-livestock integration today wherein livestock graze on crop stubble after harvest and the fields benefit from the resulting manure deposits.

Third, while cotton was originally domesticated in Egypt, Kriger (2005) estimates that cotton was likely present in Mali from the tenth century AD,[6] having arrived via the Trans-Saharan trade. In subsequent centuries, Arab travellers report cotton being widely grown in small gardens by Malians as a perennial crop (in the form of bushes) for artisanal cloth production. In this time period, many agricultural households would have spun and woven their own cotton textiles during the long dry season when farm work was minimal. In fact, visitors in the early colonial period report seeing artisanal looms everywhere in rural areas in the dry season (Forbes 1933). As such, cotton production in this time period was small scale and locally oriented.

In sum, agriculture in the precolonial period was highly adapted to local environments, from the floodplains of the inner Niger delta, to the savanna grasslands of the south. There were also multiple livelihood strategies (farming, herding and fishing) that co-existed and exploited different ecological niches. In many ways, these systems were based on farmers and producers sound understanding of agroecology, which helps us understand why they were so productive. Cotton was also integrated into these production systems as a garden crop and did not compete with food crop production for labour. While large organized empires existed in the precolonial period (such as the Empires of Ghana, Mali and Songhay), and extracted surplus production in the form of tribute, they did not interfere with the manner in which farmers went about producing food.

Malian agri-food systems in the colonial era and the emergence of cotton as a cash crop

Ever since the American Civil War disrupted cotton exports from the United States and fomented social unrest among laid off textile workers in Europe (Henderson 1933), France had been aiming to develop alternative sources of cotton in its colonies (Roberts 1996). The French saw Mali as a rich agricultural

---

6. The assertion that cotton was widespread in the region from the tenth century is based on three types of evidence: (1) the discovery of cotton seeds in archaeological sites in the upper Niger bend; (2) archaeological finds of ceramic spindle whorls – used to weight spindles when hand spinning yarn; (3) the discovery of cotton cloth fragments in burial sites on the Bandiagara escarpment that date to this period (preserved because of exceptionally dry conditions) (Kriger 2005).

zone strategically located along the Niger River. At the heart of their agricultural development efforts in French Soudan (now Mali) was the Office du Niger dam and irrigation project (see Figure 4.2), spearheaded by Emile Bélime, a civil engineer who believed he could create a project that rivalled the British irrigation schemes in Egypt (Bélime 1925, 1941). Bélime drew on the work of French agronomists who had done tests with Egyptian cotton on the Senegal River in the 1860s as well as studies by American agronomist R. H. Forbes[7] (Forbes 1933). Bélime successfully sold the Office du Niger idea to the French Minister for the Colonies, Albert Sarrault, who put pressure on the Governor General of French West Africa, Jules Carde, to support the project (Becker 1994). The negotiations with Carde led rice to be added to the scheme (at Carde's insistence) and Bélime was named as director (of Service Générale des Textiles et de l'Hydraulique) to oversee the initiative (Schreyger 1984). Interestingly, with the addition of rice to the initiative, Bélime now began to market it as a bulwark in "the fight against famine" (Herbart 1939), giving insight into the cynical way that hunger reduction could be used as a banner to achieve other objectives.

This was a massive infrastructure development initiative involving a dam on the Niger River at Markala, a series of irrigation canals, and a dream of plantation style cotton and rice production. Central to this initiative was the idea of European-managed agriculture, which stood in contrast to the peasant managed model (Roberts 1996). Launched in the 1920s, the scheme was to produce cotton for the French textile industry and rice as a food crop for the region, including rice for rural households in Senegal that were increasingly being asked to produce peanuts for export markets (Van Beusekom 2000). The Casamance area in southern Senegal was at the centre of that country's so-called peanut bowl and many Malians temporarily or permanently migrated there to work in the peanut fields. In fact, I remember talking to old men in southern Mali who had literally walked to Senegal, worked there for several years, and then walked back home (a distance of 1,100 kilometres one way if I measure it using Google Maps today).

The Office du Niger project ran into troubles from the start, including social, economic and environmental challenges. First, the area lacked sufficient labour resources. In an attempt to address this problem, the French initially used forced or minimally paid labour to construct the canals. The French then obligated some of the surrounding villages to relocate into the irrigated areas and rates of malnutrition began to climb. As Becker (1994) describes, malnutrition mounted

---

7. Forbes was the chief agronomist at an agricultural research station in Tucson associated with the University of Arizona. He made several trips to West Africa (related to his interests in dryland agriculture) and published articles in the *Geographical Review* (1933, 1943). The fact that the flagship journal of the American Geographical Society was publishing these articles suggests that the discipline of geography was also implicated in the colonial process.

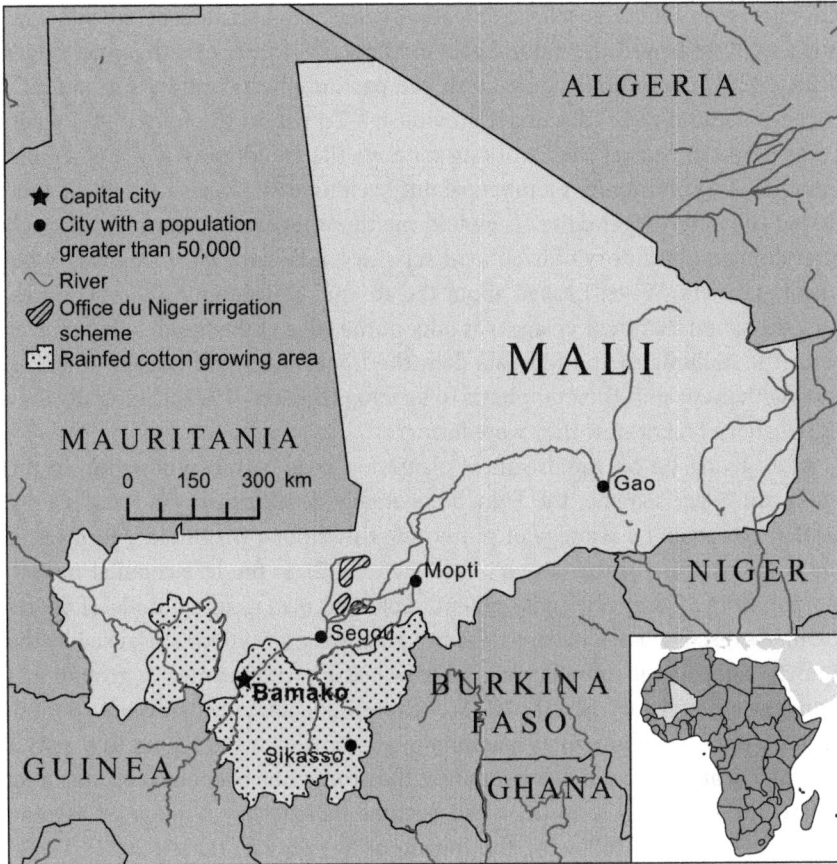

Figure 4.2 Location of Office du Niger and southern cotton growing areas
Cartography by Kellen Chenoweth, Macalester College.
Sources: World Agroforesty Center 2014; Esri Africa 2018; Humanitarian Data Exchange 2019, 2023; Africa Albers Equal-area Conic Projection.

as peoples' diets changed from diverse meals in their old villages made up of sorghum, millet, cowpeas, Bambara groundnuts and tubers, to a high starch and low protein diet of rice and a few vegetables. To solve the labour problem, the French eventually imported Mossi labourers from neighbouring Burkina Faso to farm in the scheme. Second, farmers involved in the scheme also could not produce cotton at a price that was competitive internationally, eventually leading the French to stop prioritizing cotton in the area. Third, multiple environmental problems would emerge. Poor drainage in the irrigated fields led to salinization, or the build-up of salts in the soil following the evaporation of irrigation waters. Furthermore, the dam at Markala, and substantial water diversion into Office du Niger irrigation canals, had significant downstream impacts. It destroyed the natural ebb and flow of water across the inner Niger delta, greatly compromising rice, fish and pasture production in this naturally rich agricultural zone (discussed previously). To wit, in the mid-1990s, when I was based in central Mali working for Save the Children (UK), I remember interviewing community members about livelihood strategies in a village bordering the inner Niger delta. They told me that they were fisherfolk, although after doing an inventory of livelihood activities, it became clear that they were mainly farmers. When I asked about the absence of fishing in their activities, they explained that their village was once at the edge of the floodplain, but ever since the building of the Markala dam the floodwaters had ceased to arrive, forcing them to shift their emphasis to farming (Moseley 1995). Culturally they were fisherfolk, but now they were farmers.

After giving up on the dream of plantation-style cotton production in the Office du Niger scheme, the French eventually developed upland varieties of rainfed cotton and a strategy of promoting this among smallholder farmers in the southern part of the country (see Figure 4.2) in the late colonial period, shifting from a European management to peasant management model (Roberts 1996). As discussed in Chapter 2, the head tax was a key policy stick used by the French to encourage more farmers to grow cotton, with peanuts or groundnuts being the other major cash crop the French were encouraging in this time period. Peanuts were being grown as a commercial crop across the region to supply a surging European demand for peanuts in the production of cooking oil, machine grease and especially soap (that had become increasingly popular as hygiene standards improved following the Industrial Revolution) (Lewis 2022). While peanut production would eventually decline due to a variety of factors, including soil degradation, changing global tastes for vegetable oils and an aflatoxin scare, cotton production would steadily increase from the 1950s. A key challenge for the French were local weavers who were often willing to pay more for cotton than the export price. This led to a certain amount of increased cotton production being diverted to local cotton markets (which was enormously frustrating for the French) (Roberts 1996). This problem was eventually solved by flooding local markets with cheap imported cotton textiles, which drove many local weavers out of business. That said, one could still see local weavers working outdoors in the dry season in the early 2000s.[8] The sale of cheap, second-hand clothes (often shipped to Africa from the Global North as donations) have probably put the final nails in the coffin of local textile production (Brooks & Simon 2012).

---

8. I observed remnants of this in the 1980s, 1990s and 2000s, that is, older men weaving cloth in the dry season in small villages (see https://youtu.be/KgjHc5BQ0ao). Others have remarked on this as well (Klein 1998).

As export-oriented cotton production took root in southern Mali in the 1950s, a number of important changes began to occur. First, cotton was framed or positioned by the French as a crop that men would grow, marginalizing the many women who historically grew cotton in their gardens. This was not only done discursively, but all of the cotton companies extension activities exclusively targeted men. This further tilted the balance of power in households as men earned more money. Second, the French incentivized some farmers to grow cotton by giving them mould board ploughs. Ploughs had been somewhat uncommon in southern Mali because of their expense and the tsetse fly problem in more humid areas that made it challenging to keep draft animals. In distributing ploughs, the French tended to give them to the largest, wealthiest and more influential households. This unequal distribution of ploughs had profound implications in the region given the nature of the land tenure system. As discussed in Chapter 1, most of Mali's rural areas have and do operate under a common property system. Under such a system, the first family in a village to clear a patch of land for farming held the usufruct or use rights to the parcel. These land rights may be passed down from one generation to the next but not bought and sold like private property. Ploughs are a labour-saving device, and by distributing them unevenly, the French essentially enabled wealthier households to clear and farm more land, leading to a widening gap between the rich and poor in terms of land rights. The other thing the plough did, with its pattern of long linear furrows, was to foster more soil erosion. Despite these emerging social and environmental problems associated with cotton in the late colonial period, cotton was firmly established as a cash crop at the time of independence (Moseley 2001).

The major take-away from this period is that French agronomists and engineers were deeply involved in colonization initiatives in Mali (then French Soudan), demonstrating the links between intellectual and political decolonization. They spent millions developing the Office du Niger irrigated cotton scheme, not to mention the countless West Africans who lost their lives or were malnourished in the process. Their dream was European-managed agriculture, engineered landscapes and imported seed technologies. While the project was fundamentally about producing cotton for the French, as well as rice to feed dryland farmers producing peanuts, it was frequently framed as a hunger alleviation effort from its earliest days. When this approach to cotton production failed spectacularly because of economic difficulties and environmental problems (not to mention social costs that did not figure into their calculus), the French switched to a peasant managed model of cotton production in the southern part of the country, and a carrot and stick approach to motivate farmers to produce cotton for export. In the process, cotton moved from not competing with food production, and one of many crops in a farming system, to a central focus of labour, eventually detracting from food production. As I will show, these trends will only grow worse in the postcolonial period.

Cotton becomes king: agricultural development after independence

While Mali tried to develop a more diversified economy and modest industrial base following independence in 1960, these attempts were fairly short-lived. Mali's socialist leaning first president, Modibo Keita, also experimented with collective farm production in the early years of his presidency (inspired by Soviet and Chinese models) (Painter 1978). Following poor results and farmer resistance, the government eventually and quietly signed a ten-year agreement in 1964 with the French Company for Textile Development (CFDT) to manage and further develop the country's cotton production. The CFDT was half private and half owned by the French government. This agreement gave the CFDT monopoly control over cotton production in the southern part of the country. The cotton value chain system (or *filière coton*) was developed in this period, a vertically integrated system wherein the CFDT provided inputs to farmers on credit, was the guaranteed buyer for the crop at harvest, fixed the price of cotton at the start of the system, and then ginned (a mechanical process that separates cotton fibres from their seeds) and sold the crop on international markets. The CFDT also invested in seed development, promoted the use of inorganic fertilizers and pesticides, and further developed the use of animal traction. In 1974, the government nationalized the cotton sector. They created the Compagnie malienne pour le développement des textiles or Malian Textile Development Company (CMDT), with the government owning 60 per cent and the CFDT owning the other 40 per cent (Theriault & Sterns 2012).

By the end of the 1960s, the joy of independence had faded, Mali's first president, Modibo Keita, had been ousted from power in a military coup, and Mali's economy was stagnating. Over the 1970s, corruption emerged as a growing problem and Mali's government and state-run companies became bloated through patronage hiring. By the early 1980s, with the possibility that Mali might default to its borrowers, the country's military leader Moussa Traoré[9] was forced to accept structural adjustment loans from the World Bank and IMF. In exchange for these loans, Troaré's regime was obligated to undertake a set of neoliberal

---

9. I superficially met Moussa Troaré once at the end of a running race in Bamako in 1987. He was in a viewing stand at the finish line and I was asked to shake his hand after I spoke to some reporters. I had a dim view of Traoré given that he ran a police state that imprisoned dissenters and allowed corruption to take root in the country. As such, I said the customary greetings in Bamanankan and then quickly departed.

economic reforms, including reductions in the size of government, the privatization of state-run firms, the removal of subsidies and tariffs, and an emphasis on the export of goods for which Mali was deemed to have a comparative advantage. This last condition set the tone for Mali's growing emphasis on cotton production.

Cotton production is now well ensconced in the southern part of the Mali, with production figures steadily climbing over the past 40 years (with some dramatic ups and downs along the way, such as a cotton farmers' strike in 1999 and defections due to low producer prices in 2007) (see Figure 4.3). This rise in cotton production has been celebrated as an agricultural development and food security success, especially by western donors such as CIRAD and USAID (Fok 2000; Tefft 2004). Spatially, the original French efforts aimed at promoting rainfed, export-oriented cotton in the 1950s in the old cotton basin, with the town of Koutiala as its main hub (zone A in Figure 4.4). Then, in the 1970s, the CMDT began to focus on an area south of Bamako known as the new cotton basin (zone B in Figure 4.4). Lastly, in the 1990s, the CMDT focused on developing production in an area known as the cotton frontier with the town of Kita as its main hub (zone C in Figure 4.4). My previous research in old and new cotton basins showed that intensive cotton cultivation in southern Mali led to soil degradation, especially a problem known as soil acidification (Moseley 2005b, 2008b). Soil acidification occurs with repeated use of inorganic fertilizer over a prolonged period with limited use of organic matter (read manure and compost) to maintain soil health. This problem led to declining yields, and it was one of the factors pushing cotton production into new areas of southern Mali. As yields declined in the old cotton

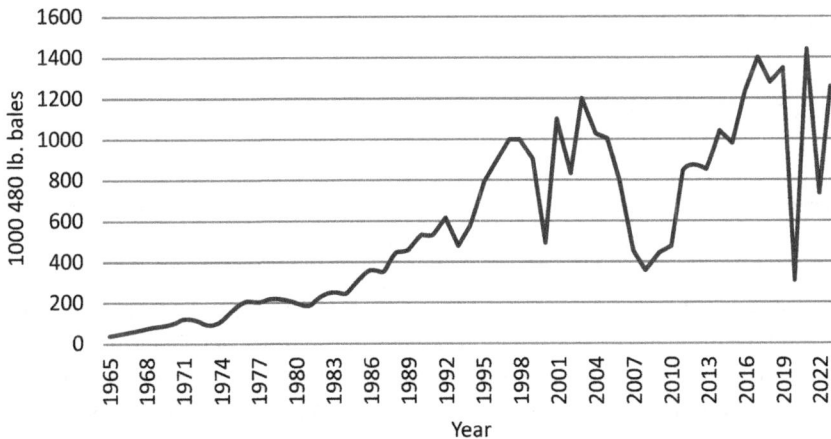

Figure 4.3 Mali's cotton production 1965–2023
Source: Graph by author based on US Department of Agriculture Data (2023).

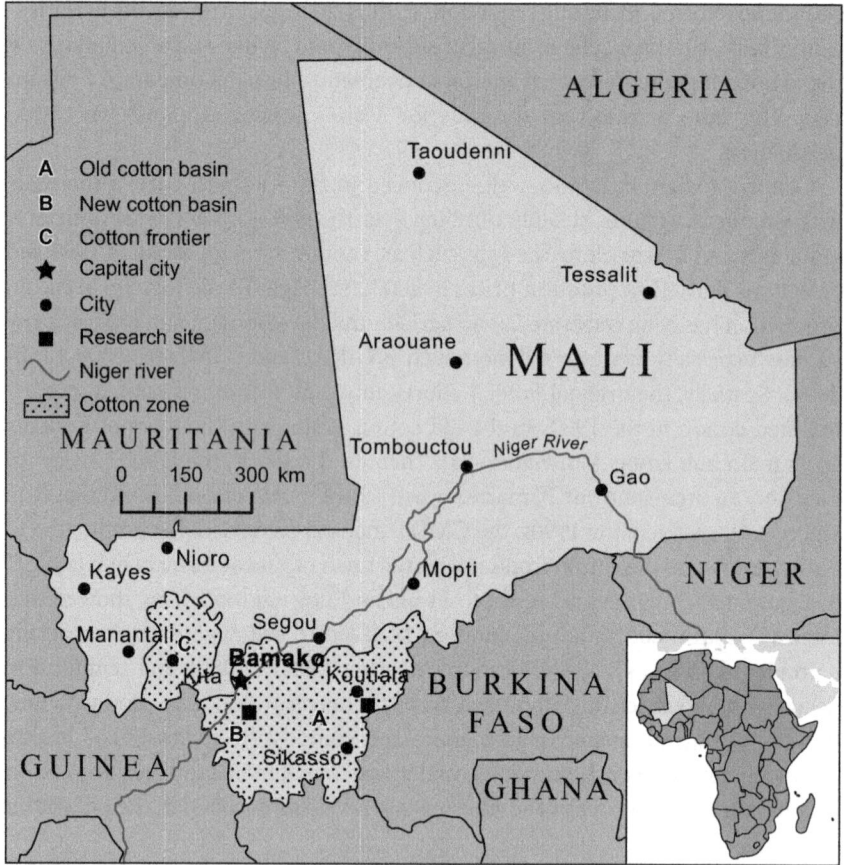

Figure 4.4 Mali's cotton zones
Cartography by Kellen Chenoweth, Macalester College.
Sources: World Agroforesty Center 2014; Esri Africa 2018; Humanitarian Data Exchange 2019, 2023; Africa Albers Equal-area Conic Projection.

basin, the CMDT moved into the new cotton basin in order to maintain production, and then eventually into the cotton frontier for the same reasons.

In addition to soil degradation, cotton production is associated with heavy use of pesticides and herbicides. My research in the early 2000s showed that those farmers growing more cotton tended to use more pesticides per hectare than those who grew less or no cotton (Moseley 2005b). Heavy pesticide use is a problem for human health and the environment. Most cotton pesticides in Mali are applied using backpack sprayers and little to no protective gear. Furthermore, most safety information on labels is published in English, French and/or German languages that most Malian farmers do not read. There is also research from neighbouring Burkina Faso (which also grows a lot of cotton) suggesting that pesticide runoff into nearby standing water has interacted with the evolutionary trajectory of the Anopheles mosquito (which carries malaria) and the development of new strains of malaria that were resistant to common malaria medications (Dabiré *et al.* 2012). Furthermore, since the early 2000s, there has also been a considerable uptick in the use of herbicides that is well documented in Mali (Haggblade *et al.* 2017a). The key ingredients in the most commonly used herbicides are glyphosate (a suspected carcinogen) and atrazine (a known hormone disruptor). Further aspects of the health concerns related to growing herbicide use will be discussed in Chapter 5.

In theory, one of the main factors driving cotton production in Mali has been its purchase price as well as the cost of related inputs.[10] Nonetheless, in many years the margins on cotton production are incredibly small. As such, access to credit and inputs has been important for sustaining cotton production through less profitable years. These inputs are often used on food crops which also benefit from the machinery (such as ploughs) that cotton cultivation helps farmers acquire. As a result, even when the margins on cotton are low, and its cultivation damages the soil (which many farmers recognize), farmers may still grow the crop for these other reasons.

In addition to the environmental challenges of the soil degradation and chemical use related to cotton production, and the related expansion of cotton into new areas in order to maintain output, there were social and food security implications (which are deeply intertwined) that accompanied the growing emphasis focus on cotton. For starters, cotton is a very labour-intensive crop, especially at key junctures in the agricultural season such as planting, weeding and harvesting. It is at these moments when male heads of household typically call on all household members to work in the family's cotton fields (see Figure 4.5). While this seems reasonable enough at first blush, the problem is that it pulls women away from labouring on their own food crop fields. As women are responsible for growing the "sauce" crops (see footnote) in the Malian context (various leafy greens, peanuts, cowpeas, hibiscus),[11] women's labour being diverted to cotton production means that the nutrient dense part of meals may be compromised.

Cotton farming has produced profits for male heads of household that were often invested in cattle (which operated as a store for wealth). As a consequence,

---

10. One of the main reasons Malian cotton production rose in 2022 (see Figure 4.3) is that the government chose to subsidize inputs. Such subsidies have been discouraged by the World Bank and the IMF since the 1980s.

11. Certain aspects of agriculture in southern Mali are gendered (and across the larger Mande speaking world in West Africa) where men are responsible for providing the staple grains and women supply the ingredients of the sauce that accompanies the main starch in a meal. Crops that contribute to the sauce are sometimes referred to as sauce crops.

Figure 4.5 A family working in the fields during the cotton harvest, late 1980s
Source: Photo by author.

the cattle herd in southern Mali has grown immensely over the past several decades and now surpasses that of the semi-arid Sahel zone where more cattle used to be kept. These cattle were often kept on common areas near villages. The resulting compaction of soils led to increased water runoff during rainfall events and a growing erosion problem in neighbouring fields. As such, everyone in the community bore the costs of the cattle investments of the wealthy (Moseley 2008b).

The last and most serious problem is that of growing malnutrition in the southern part of the country (alluded to at the start of this chapter). Originally identified as the Sikasso Paradox by French researchers, Dury and Bocoum (2012) argued that this was related to household structure in the region. That is, large extended families, headed by older males who re-invested cotton proceeds in production rather than nutritious food for children. They speculated that if younger parents controlled the purse springs, then different decisions about diet might be made. While this might be true, the increasingly monetized nature of the food economy also suggests that food expenditures now compete with capital expenditures in a way that they did not when healthy food was produced in-situ by the family rather than purchased. The other contributing factor to malnutrition is the explosion of maize production in the region that has grown in tandem with cotton production (Laris & Foltz 2014). Cotton has normalized the use of inorganic fertilizer in southern Mali (if not created a dependency), and maize has been bred to be much more responsive to fertilizer inputs. As such, maize has now supplanted millet and sorghum as the main staple food grain in much of southern Mali. Maize is less nutritious than the traditional small grains (lower protein) and less likely to be intercropped with other nutrient dense crops like cowpeas.

Conclusion

Mali's postcolonial period has been characterized by the growing importance of cotton as a commodity crop for export. This did not happen by accident. The French had begun pushing export-oriented cotton in the colonial period and did much to destroy the local weaving industry. Then a combination of Malian government policy, French agronomic expertise and pressure from the World Bank and IMF created a system that incentivized Malian farmers to keep producing more cotton even when it was bad for the environment and food security. Of course, farmers occasionally exerted their agency, with a notable strike that occurred in 2000 and a good deal of passive resistance in 2007–08, but good old-fashioned political ecology analysis suggests that the imperatives of the government and international policy-makers were privileging commodity production over food security, even if the former might cynically be promoted as the latter. That said, the aforementioned strike and resistance do offer up interesting unplanned experiments in an alternative, food first approach, and a reason for hope that will be explored in Chapter 8.

5

PLANNED AND UNPLANNED AGRICULTURAL CALAMITIES IN BURKINA FASO: THE NEW GREEN REVOLUTION FOR AFRICA AND PROLIFERATING HERBICIDES

In 2016 I started working in Burkina Faso because the insecurity in Mali had become too great to bring my students there as research assistants. Of course, the benefit of working in southwestern Burkina Faso was that it had linguistic and cultural similarities to the southern area of Mali that I knew well. Nonetheless, I was pleasantly surprised to learn that the chief of a Mossi village where I was doing research spoke Bamanankan or Dioula, not always the case for the Mossi who speak Mòoré and are the largest ethnic group in Burkina Faso.[1] What was even more interesting was why this chief spoke Dioula so well. The Mossi come from the central plateau region of Burkina Faso, a densely populated and economically depressed area. During the colonial era, the Mossi were often recruited by the French to work on agricultural schemes because of their strong work ethic (Filipovich 2001). As a consequence, the chief's family had been moved to the Office du Niger project in Mali (discussed in the previous chapter) in the colonial period to work in the rice and cotton fields. Here he grew up and learned Bamanankan. Given his family's rice production expertise, they were then resettled in the 1980s back to Burkina Faso along the Mouhoun River (also known as the Black Volta) in a new area where they were trying to promote rice production following spraying for the black fly and river blindness control (McMillan 1995). I would chat with the chief many times over the next five years as we repeatedly visited his village to interview women

---

1. Other groups in Burkina are more likely to speak Dioula than the Mossi, such as the Bobo, Gurunsi and Senufo in the southwestern part of the country. Dioula literally means trading language as it was a lingua franca in the region prior to French. Dioula is a second language for many people and is a somewhat simplified version of Bambara and Malinke, which are mutually intelligible Mande languages and the mother tongues of these ethnic groups in Mali and Guinea. The second largest city in Burkina Faso is Bobo Dioulasso in the southwest, which literally translates to the house (*so*) of the Bobo and Dioula as the city was established by members of the Bobo ethnic group and Dioula speaking Muslim traders.

farmers about the nutritional impacts of a New Green Revolution for Africa (GR4A) rice project.[2]

Burkina Faso has an agricultural development trajectory that is similar in many ways to Mali with its emphasis on cotton. However, after sharing some basic background information on Burkina Faso, in this chapter I focus on two other facets of agrarian change in the postcolonial period with deep involvement of international donors and the private sector: the GR4A with its focus on improved inputs and commercialization, and a disconcerting proliferation of herbicides (which is not disconnected from Burkina Faso's history of cotton production).

Understanding Burkina Faso

Burkina Faso has a population of 22 million people and is 274,220 km² in size, giving it 22 per cent of the land area of neighbouring Mali with roughly the same size population. The country is 70 per cent rural and 30 per cent urban with 73 per cent of people dependent on agriculture. Ouagadougou (the capital city in the central area) and Bobo Dioulasso (in southwestern Burkina Faso) are the two most important cities. Burkina Faso has similar climate zones to neighbouring Mali, excepting the Sahara Desert to the north (which does not dip down into Burkina Faso) (see map of climate zones in Figure 4.1 in previous chapter). The northern parts of Burkina Faso are considered to be part of the Sahel or semi-arid savanna with less than 600 millimetres (mm) of rainfall per year. The southern parts are humid or Guinea savanna, which is sometimes split into two zones based on rainfall (600–900 mm per annum and 900–1,200 mm per annum). While animal husbandry and agropastoralism dominate in the Sahel region, farming is more common in the more humid savannas. The rainy season is three to five months in length depending on latitude, progressively decreasing from north to south (May/June–September/October). The traditional food crops were millet, sorghum, cowpeas and peanuts. Maize has now become much more dominant in the more humid southwest as it (like in Mali) is increasingly grown in rotation with cotton. The country's former name, Upper Volta, refers to three branches of the Volta River that extend up into the country from Ghana: the Black, White and Red Volta. The Mossi are the largest ethnic group, followed by the Fulani. The country's major exports are gold and cotton (Echenberg et al. 2023).

---

2. See https://youtu.be/MU3SREuHuxo, where I am chatting with the chief in June 2019 in a brief video in English, French and Dioula.

In the precolonial era, the area was often associated with the Mossi Kingdoms that flourished in the fifteenth to eighteenth centuries, centred around Ouagadougou (and resisted incorporation into the Ghana, Mali or Songhay Empires discussed in the previous chapter). The area succumbed to the French in the late nineteenth century (1898) and was originally governed as part of the large French Colony of Upper Senegal and Niger between 1904 and 1919. Upper Volta was spun off as a separate colony in 1919 in part because of an indigenous uprising between 1915 and 1917 known as the Volta-Bani War (Şaul & Royer 2001). While not widely known or studied, historians argue that this was one of the most successful anticolonial struggles on the continent. It took place in communities across a wide area in contemporary southeastern Mali and south-western Burkina Faso and was centred around an area just north of the city of Bobo Dioulasso. The struggle started during the rainy season of 1915 when a collection of villagers gathered expressing discontent with the French occupation, but especially against the conscription of Africans to fight on behalf of the French in the First World War. The insurgents also recognized that the French were particularly weak at this moment as their resources were tied down in the European war. At the height of the conflict in 1916 there were 15,000–20,000 local men fighting against the French on multiple fronts. While the French eventually put down the insurgents, jailed and executed their leaders, and mowed down civilians with machine guns, there was still active resistance until 1917. The hiving off of Upper Volta in 1919 as a separate colony was in part a tactic to split apart the areas involved in the resistance, dividing them between Sudan (now Mali) and Upper Volta (Şaul & Royer 2001).

Upper Volta would eventually gain independence from the French in 1960. Maurice Yaméogo served as the first president. Yaméogo was a somewhat unlikely leader, quickly rising to power after the untimely death of Ouezzin Coulibaly[3] who was president of the first governing council of Upper Volta in the transitional years leading up to independence and a driving force behind the anticolonial struggle. Yaméogo served as his second in command. Unfortunately, Yaméogo had autocratic tendencies and would be ousted in a coup in 1966. Lieutenant Colonel Sangoulé Lamizana would then unremarkably rule the country until 1980, at which time there were worker strikes and then another coup. After several transitions, Captain Thomas Sanakara came to power, leading to

---

3. Ouezzen Coulibaly was a close friend of Mali's first president, Modibo Keita. In Coulibaly's honour, Keita (with Russian financing), created a village named Ouezzendougou with streets on a grid pattern (not typical in traditional villages) just southwest of the Malian capital city, Bamako. I lived in Ouezzendougou (which means the town of Ouezzen) for four months in 1987 when I was training to be a Peace Corps volunteer.

one of the more interesting periods in Burkina Faso's postcolonial presidential history (Englebert 1996).

While much could be said about Sankara, perhaps what is most relevant to this chapter is that he and his successor Blaise Compaoré are often associated with two very different agricultural development legacies in Burkina Faso. Ruling between 1983–87, Sankara was a Marxist inspired leader and a huge proponent of self-sufficiency and anti-colonialism. He changed the name of the country from Upper Volta (Haute Volta) to Burkina Faso (a combination of the Mossi word for honest man and the Dioula word for country). He also lived simply and by example, driving his own small sedan (a Citroen known as a *deux cheveaux* or two horses for its two-horsepower engine) to work and back (remarkable for a time when many African leaders lived ostentatiously and beyond the means of their countries). More importantly, he sought to foster a small industrial base in the country, supported the rights of women, and essentially promoted agroecology via his work with agroforestry and organic agriculture (Dembele 2013). Some aspects of Sankara's agroecological legacy remained alive in civil society and non-governmental organizations, for example, the Six-S movement (Lecompte & Krishna 1997). In contrast, Blaise Compaoré, who murdered and deposed Sankara in a coup, ruled for over 25 years (1987–2014) and was a big proponent of more conventional agricultural development approaches. He adopted structural adjustment measures (discussed in Chapter 2), oversaw the rise and expansion of cotton production in the country, and allowed for the use of Bt cotton (a form of GMO cotton) (Dowd 2008).

Compaoré left power after popular uprisings in 2014. The country held democratic elections in late 2015, with Roch Kaboré becoming the first democratically elected president in nearly 50 years and ruling until January 2022 when he was ousted in a coup. Kaboré's tenure is significant because of the influx of foreign assistance during this period, including much of which was agriculture and food security related. The GR4A initiative came to the fore in this time period as well as some other unexpected legacies from cotton production from the Compaoré years.

Redux of a square peg solution to a round hole problem: the new green revolution for Africa (GR4A)

As discussed in Chapter 2, the first Green Revolution was launched in the 1950s and 1960s to bring industrial food production techniques to the Global South. At that time, there were growing fears in the Global North regarding population growth and food shortages in the Global South (e.g. Ehrlich 1968) and the Green Revolution was seen as a productionist solution to this problem. This was also the era of the Cold War and western countries were worried that hunger-related social unrest would foster socialism (Patel 2013). In fact, the term "Green Revolution" was coined by former US Agency of International Development (USAID) administrator William S. Gaud who saw it as an antidote to communism. In a 1968 speech, he said: "This and other developments in the field of agriculture contain the makings of a new revolution. It is not a violent Red Revolution like that of the Soviets. I call it the Green Revolution" (quoted in Dalrymple 1979: 723–4). This is arguably one of the more explicit examples of how agricultural development efforts were inflected with politics. Furthermore, agronomist Norman Borlaug[4] would eventually win the Nobel Peace Prize for his work breeding dwarf wheat varieties and "saving the world from hunger", a legacy that is continuously cultivated and celebrated with generous support from corporate agriculture. While the Green Revolution is widely fêted in conventional agricultural circles as a success given improved yields, this effort had some serious shortcomings (Moseley 2015d). First and foremost, it only addressed one of the six dimensions of food security discussed in Chapter 3 (availability via increased production). It also led to increasing social stratification as only wealthier farmers could adopt the approach. There were also a number of environmental problems, including pesticide resistance[5] and issues related to river dams and water diversion (Moseley 2017a). There was also a belief that the first Green Revolution largely bypassed Africa because it did not focus on crops widely grown on the continent (but rather wheat and floodplain rice).

Proponents of a second wave of the Green Revolution focused on Africa (known as the New Green Revolution for Africa (GR4A)) were able to capitalize on the 2007–08 global food crisis and push their agenda forward. The situation in Burkina Faso made the country's government particularly receptive to such an endeavour. Like many countries across West Africa, as Burkina Faso's population urbanized, it shifted away from coarse grains like millet, sorghum and maize as major foodstuffs, and increasingly consumed more rice. My surveys with household cooks (all women) in the urban areas of neighbouring Mali showed that they preferred rice because it was quicker and easier to prepare, socially it had more status as a food stuff, and it expanded when cooked, leaving people feeling sated (Moseley 2011). The problem for Burkina Faso was that it only produced 25 per cent of its

---

4. Norman Borlaug's legacy continues to be celebrated in a variety of ways. A statue of him stands in the US Capitol rotunda (one of two statues representing personages from Iowa) and the World Food Prize is closely associated with his legacy (and is bestowed annually in Borlaug's home state of Iowa). Both the Norman Borlaug Foundation and the World Food Prize Foundation benefit from generous corporate agriculture support.

5. Insect communities develop resistance to pesticides over time. No pesticide is 100 per cent effective. If a few individuals survive, they will reproduce. Eventually and gradually the insect population will be replaced by individuals with greater tolerance and the farmer will need to apply more and more pesticide to get the same effect (creating the problem of pesticide resistance).

own rice and imported the rest. This became a major problem during the 2007–08 global food crisis when rice prices went up 100 per cent, leading to social unrest in Ouagadougou and Bobo Dioulasso, not to mention other cities across West Africa (Engels 2015). In many ways, this was a classic food access problem, as rising food prices, especially for staple grains, means that the poorest of the poor are unable to access adequate calories unless cheaper substitutes are available.

The GR4A was similar in some ways to its first iteration (with a focus on production and the use of improved seeds and inputs) but also different. The differences included a focus on African crops, women farmers, greater involvement of the private sector and an explicit attempt to connect farmers to broader markets (and these connections were often conceptualized as a value chain). It was further argued that increased profits would lead to gains in nutrition, especially if women were engaged in these projects. One of the major funding vehicles for GR4A has been the Gates Foundation supported Alliance for a Green Revolution in Africa (or AGRA).

My collaborators and I examined an AGRA supported rice commercialization project in southwestern Burkina Faso over a five-year period from 2016 to 2020 (Moseley & Ouedraogo 2022). In an effort to reduce Burkina's dependence on imported rice, the project aimed to bolster domestic production of new rice varieties that appealed to the tastes of Burkina's urban consumers. Substantial funding went into training and supporting Burkinabè seed scientists who developed these new seed varieties. Grants were also provided to support the development of private sector seed companies, such as Neema Agricole du Faso SA (NAFASO) and Faso Agriculture (FAGRI), that replicated and marketed the new certified seed. The project then lined up rice millers who would sell the improved rice seed to farmers in project villages, purchase the harvest, mill the seed, and then have it packaged and branded as Burkina Rice (*Le Riz du Burkina*) before it went to retail shops for sale. The project also worked to market Burkina Rice to consumers (see roadside advertising billboard in Figure 5.1).

At the village level, the project worked in villages where other development organizations had levelled fields in traditional rice growing areas, divided these into quarter hectare plots and used village labour to construct micro-dikes for the partial control of water that naturally flowed into the area during the rainy season, either because the fields were located along a river or in a low-lying area. At least a third of the participants in these schemes were women. See a map of our study villages in Figure 5.2, including project and control villages.

Unfortunately, this project has not improved the food security and nutrition of participating female farmers. Our sample was split between women rice farmers who participated in the project and those who did not (our control group) (see Table 5.1 with descriptive statistics). We measured food security and nutrition quality using three different survey instruments: the household food insecurity

Figure 5.1 Billboard on the main highway between Ouagadougou and Bobo Dioulasso extolling the virtues of Burkina rice (*le riz du Burkina*): good taste, healthy, natural, fresh and nutritious
Source: Photo by author.

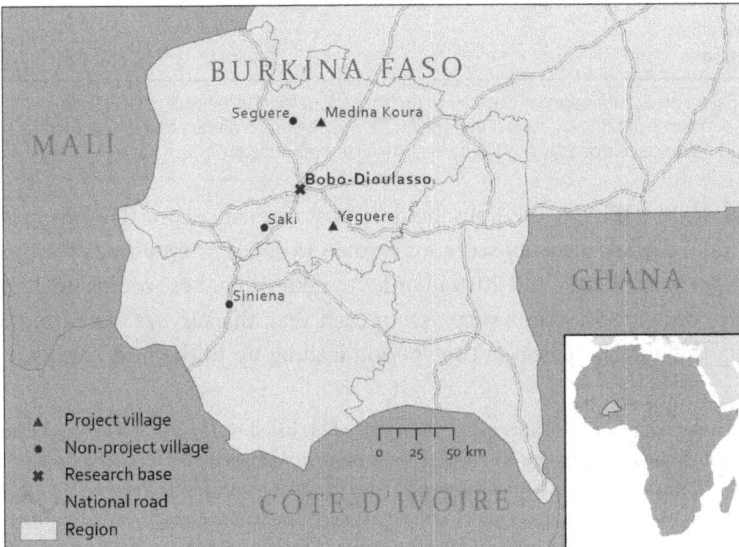

Figure 5.2 Map of research communities in southwestern Burkina Faso
Cartography by Kellen Chenoweth, Macalester College.
Sources: ESRI Africa 2018; Humanitarian Data Exchange 2022, 2023; World Bank 2023g; Africa Albers Equal-area Conic Projection.

Table 5.1 Food security and nutrition quality measures for women farmers and their families in a GR4A rice commercialization project, Burkina Faso

	HDDS		MDDW		HFIAS	
	Non-participants	Participants	Non-participants	Participants	Non-participants	Participants
Index score, mean (sd)	6.1 (1.3)	5.7 (1.2)	4.5 (1.1)	4.3 (1.0)	3.1 (0.8)	3.2 (0.9)
Interview phase, n (%)						
1	93 (25%)	70 (29%)	93 (25%)	70 (29%)	93 (25%)	70 (29%)
2	93 (25%)	70 (29%)	93 (25%)	70 (29%)	90 (24%)	70 (29%)
3	95 (25%)	49 (20%)	95 (25%)	49 (20%)	94 (25%)	48 (20%)
4	93 (25%)	53 (22%)	93 (25%)	53 (22%)	92 (25%)	53 (22%)
Wealth tertile, n (%)						
Low	27 (29%)	29 (41%)	27 (29%)	29 (41%)	27 (29%)	29 (41%)
Middle	54 (58%)	30 (43%)	54 (58%)	30 (43%)	54 (58%)	30 (43%)
High	12 (13%)	11 (16%)	12 (13%)	11 (16%)	12 (13%)	11 (16%)
Total households (all phases)	115	76	115	76	115	76
Total interviews (all phases)	374	242	374	242	369	241

Note: Based on author's survey Data. HFIAS = household food insecurity access scale. HDDS = household dietary diversity score. MDDW = minimum dietary diversity score for women. Table developed by Matt Gunther, University of Chicago.

access scale (HFIAS), the household dietary diversity score (HDDS) and the minimum dietary diversity score for women (MDDW).[6] Questions were asked twice per year in 2017 and 2019 in order to determine these scores (in 2016 we simply conducted baseline surveys). In each year, the surveys were conducted during the hungry season (a time period leading up to the next harvest when

---

6. The household food insecurity access scale (HFIAS) is based on the answers to nine qualitative questions about the previous four weeks. These range from more moderate questions (how many days did you worry about not having enough food to eat?) to more severe (how many times in the past four weeks did you go 24 hours without eating?). The score ranges from 0 to 4, with higher scores indicating greater levels of food insecurity (Coates et al. 2007). The household dietary diversity score (HDDS) is based on a recall of all the foods household members have consumed over the past 24 hours. These foods and ingredients are then categorized according to major food groups and a score for the total number of food groups is given (with higher scores being better) (Swindale & Bilinsky 2006). While this is a measure of dietary diversity, it is also considered to be a proxy for food security. Lastly, the minimum dietary diversity score for women (MDDW)

food is scarce) and in the post-harvest period when food is more plentiful. There was no significant difference in all three measures (HFIAS, HDDS and MDDW) for food security and nutrition quality between women who participated in the rice commercialization project and those who did not.

Furthermore, if you look at individual types of foods consumed (see Figure 5.3), women who participated in the project are (by and large) not consuming more healthy food or a greater diversity of foods. One notable exception is that women who participated in the project are consuming more dark leafy greens (which is good because these tend to be rich in iron) but also fewer types of other vegetables, fish and, oils and fats. Interestingly, project participants were also consuming more sweets.

A key question is why the GR4A approach is not improving the food security and nutrition of participating women farmers? Herewith three key reasons, some of which are relevant to other African countries where this approach has been adopted. First, those who hold the power in this particular value chain are the agronomists and seed companies. They received the majority of project funding, were well networked with each other, and had designed a chain that reflected their priorities and not those of participating farmers. Having pure certified seed and rice of a certain quality was clearly the leading goal that superseded everything else (Moseley & Ouedraogo 2022). As such, many women farmers we spoke with found that growing this new rice in a particular way and selling to certain vendors was a challenge. In some cases, the miller would arrive to pick up the rice and could not pay for another few weeks. This was especially challenging for women who had debts to settle and school-related expenses for their children, so they would often side sell to another merchant to get cash more quickly (even at a cheaper price). As such, women were eventually voting with their feet when this new value chain did not work for them.

Second, private sector involvement in the value chain and GR4A approach to agriculture means that they often pilot a growing practice and then, once it is perfected, try to mass replicate it across a large number of communities as they scale up the project. The problem is that this approach often ignores local knowledge and an understanding of local specifics that is needed for something to work well (Moseley 2017b). In the case of Burkina, the issue was that land tenure regimes varied across different villages. In a village where the Mossi settled, the rice lands were controlled by men. As such, women benefitted when

---

is similarly based on a 24-hour recall of foods that a woman interviewee of reproductive age has consumed. This is scored based on the food categories most relevant to this demographic (with higher scores being better) (FAO & FHI 2016). While this is measure of dietary diversity in relation to women's minimum dietary requirements, it is considered to be a proxy for nutrition quality for the entire household (Kennedy et al. 2010).

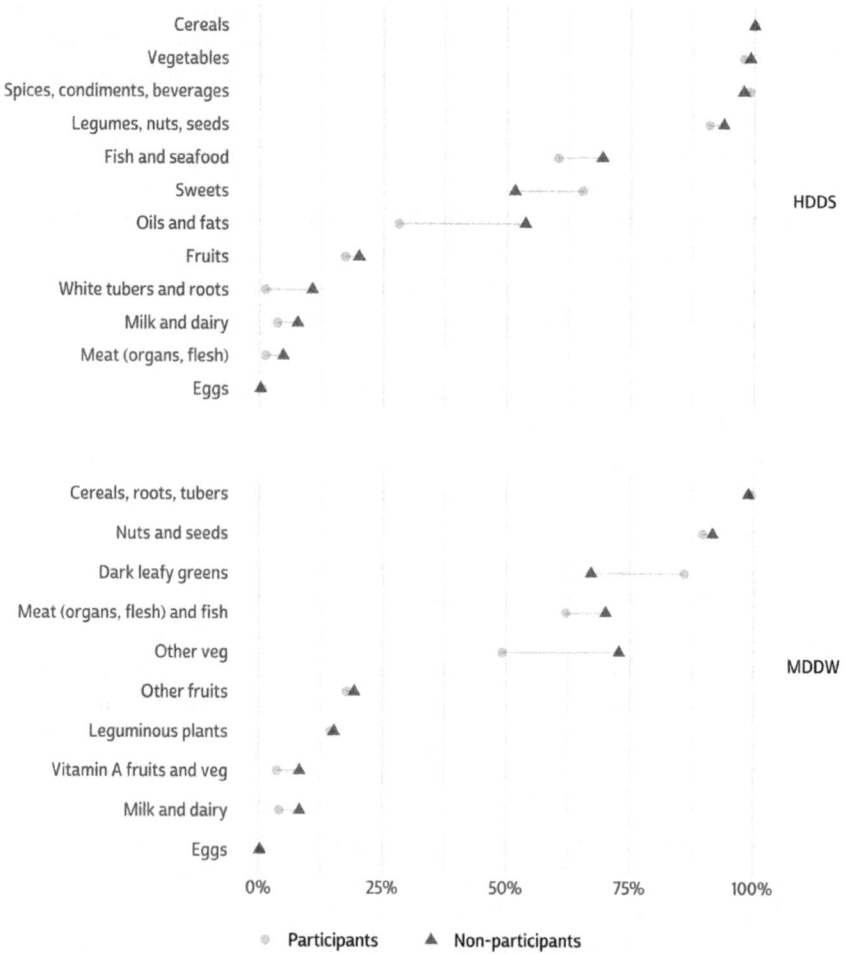

Figure 5.3 Foods consumed over past 24 hours (rice commercialization project participants versus non-participants)
Source: Based on Author Surveys. Figure created by Matt Gunther.

the project came in, improved rice lands, and then redistributed the land to men and women. However, in another community dominated by Dioula, the women controlled the rice lands. Here the women effectively lost control of some land when the project came in, improved the land, and then redistributed it to men and women (Moseley & Ouedraogo 2022). Approaches that do not account for local knowledge and variability are destined to fail.

Lastly, the idea that integrating farmers into a more monetized food economy will help them earn more money that is spent on better nutrition is questionable. When traditional, healthy foods were produced in-situ, the consumption of these items did not compete with other household priorities. Now that house-holds are producing more for the market, grow less for their own consumption and most purchase more of their own food, a new dynamic of competing house-hold expenses and priorities is created. Expenditures on nutrient dense foods are now often cut in favour of cheaper, less nutritious substitutes so that money may be spent on other priorities like school fees and medical expenses. This is leading to a nutrition transition (a switch to more processed foods) in rural areas that parallels the one that has been occurring for a much longer period of time in urban areas (Popkin 2004).

In sum, the GR4A is failing to improve the nutrition of African households. This neoproductionist approach reflects the priorities of agronomists, fails to account for the other dimensions of food security, does not adjust for local dif-ferences, and may actually be making the situation worse by contributing to a rural nutrition transition that fosters less healthy eating.

The political ecology of proliferating herbicides

In the summer of 2019, my student Eliza Pessereau and I discovered that 92 per cent of the female farmers we interviewed in relation to the GR4A rice project discussed previously were using herbicides in their fields. Most were applying these at least twice per season and some more. This was particularly striking to me as herbicide use was rare when I first started working in West Africa in the 1980s and virtually non-existent for women farmers (Moseley 1993). Now herbicides seemed to be everywhere and women used them regularly. In fact, herbicides were hard to miss as small shops and table venders in the small-est of villages would have colourful bottles of these chemicals prominently dis-played with descriptive names like *la machette* (the machete) and *la daba* (the hoe) (see Figure 5.4). In trying to understand this herbicide revolution, Eliza and I asked supplemental questions in our surveys with women farmers and other actors about this intriguing, yet disconcerting development (Moseley & Pessereau 2022).

Proliferating herbicide use is a problem because of known and unknown health consequences. In southwestern Burkina Faso, most herbicides are applied using backpack sprayers and no protective clothing. Two of the leading herbicides are glyphosate and atrazine-based. Glyphosate, which may be more familiar to some under its trademark name "Roundup", is a suspected carcinogen and hormone disruptor (Mesnage *et al.* 2015). Atrazine is a known teratogen and hormone disruptor (Hayes *et al.* 2002). Carcinogens cause cancer, teratogens interfere with the development of unborn children (particularly in the first trimester), and hormone disruptors are chemicals that mimic natural hormones and may

Figure 5.4 Village shop selling herbicides in southwest Burkina Faso
Source: Photo by author.

interfere with the sexual development of animals, including humans. Glyphosate is a broad-spectrum herbicide, killing both grasses and broad leaf weeds, while atrazine is a more selective herbicide. According to the women farmers and village-based herbicide vendors we interviewed in our surveys, glyphosate is mainly used early in the season before plants have sprouted and atrazine later in the season (Moseley & Pessereau 2022).

In the introduction to this book, I discussed Hart (2001) and Lawson's (2007) concepts of big D and little d development. Big D development referred to planned and orchestrated development projects like the GR4A initiative discussed in the first part of this chapter. In contrast, little d development refers to the economic change happening all around us that is seemingly un-coordinated. At first blush, the emergence and proliferation of herbicides use among West African farmers seems like a good example of little d development. The use of herbicides was neither clearly connected to the GR4A rice project just discussed (as the promotion of their use was not central to the initiative), nor was there some other big project pushing herbicides. That said, herbicide use did not just proliferate on its own, but rather there are historical and structural reasons for this. Political agronomy, political ecology and feminist political ecology are theoretical frameworks discussed in Chapter 3 that help us understand the rise and spread of these chemicals.

It is no accident that herbicide use in West Africa is most advanced in areas where export-oriented cotton production was promoted, including large areas of southwestern Burkina Faso that contain some of the highest cotton producing communities in the country (Haggblade *et al.* 2017a). These chemicals were initially introduced alongside cotton with their use recommended by agronomists as a labour-saving technology, indicating the role of science in their initial rollout (Grabowski & Jayne 2016; Luna 2020). Beyond this important historical precedent, the more recent explosion in herbicide use by women farmers is best explained by a political ecology approach that clarifies the links between individual farmer behaviour and shifts in broader political economy at the international, regional and national scales. Then, at the household level, a gendered or feminist political ecology lens (Rocheleau *et al.* 2013) sheds further light on the decisions that women are making when they use herbicides. Herewith a brief discussion of the important changes at the international, regional, national and household scales that fostered the rise of herbicide by among women farmers in southwest Burkina Faso.

The most significant change internationally that impacted herbicide use in Burkina Faso had to do with the agrochemical industry. The broad-spectrum herbicide glyphosate was originally developed by the American firm Monsanto (now owned by the Swiss company Bayer) in 1974 under the trademark name Roundup (Duke & Powles 2008). Glyphosate went off patent in 2000 and generic production of the herbicide increased rapidly, initially in the Global North and then in the Global South (Clapp 2021). Sales of glyphosate grew from $5.46 billion in 2012 to $8.79 billion in 2019 (Richmond 2018). Furthermore, use of the herbicide really took off in tropical countries after Global South producers (China and India most notably) were able to produce and sell it at even lower prices. In Mali, Haggblade *et al.* (2017b) found that glyphosate herbicide prices declined 35 per cent in the regional CFA currency (the same currency used in Burkina Faso), or 50 per cent in US dollars, between 2008 and 2015. Given the price consciousness of poor farmers, these declines were a major factor influencing the use of herbicides in the region, and these price reductions were directly related to changes in the global herbicide industry.

Then, at the regional and country level scales, there are two other factors that contributed to the increasing use of herbicides by Burkinabe women farmers: one regulatory and the other economic. First, after 1994 it became easier to import herbicides into West African markets. In that year regional regulations were streamlined to facilitate the importation of pesticides, including herbicides. A regional body known as the *Comité Permanent Inter-Etats de Lutte contre la Sécheresse dans le Sahel*, or CILSS, created a regulatory committee to review and certify pesticides for importation into member countries (Diarra 2015). If pesticides passed efficacy and safety tests in one of the CILSS' countries, then the chemicals were cleared for importation into all nine CILSS member states. This made the CILSS area an attractive market for pesticide manufactures (Haggblade *et al.* 2017b) because a company could target the country with the least stringent health and safety standards, and then get approval to market their product in all nine CILSS states.

Second, artisanal gold mining in Burkina Faso and surrounding countries has created labour shortages in the agricultural sector. Gold mining in the region is not a new activity but stretches back over a millennium (Kevane 2015).[7] As a result of new techniques such as cyanide leaching, modern, industrial gold mining (often run by international companies) emerged in the region in the 1980s and is a leading source of income for Burkina Faso, Mali and Ghana (Moseley 2014a; Brugger & Zanetti 2020). In addition to industrial gold mining, small-scale, artisanal mining efforts have accelerated in the past 20 years due to the availability of more afford-able technologies produced in China, such as metal detectors, excavators, crushers and sifters that make the process much more efficient (Werthmann 2017). Low-cost technology combined with higher gold prices have fostered the exodus of young people from Burkina Faso's farms to mining areas (Werthmann 2009). Young people leave the village to not only earn money but to have an income that is not controlled by their father or male head of household. In some areas of Burkina Faso, two out of every three rural households have at least one member working in the artisanal gold mining sector (Brugger & Zanetti 2020). A 2016 Government of Burkina Faso report suggested that 1 to 1.2 million people were directly or indirectly involved in the industry, of a total population of 18.6 million in that year (Assemblée Nationale 2016). Pokorny *et al.* (2019) found that artisanal gold mining was a major drain on agricultural labour in northern Burkina Faso. This finding is consistent with the overall tight rural labour market we encountered in our study area and helps one understand why it was so challenging to hire farm labour of any sort. It also helps explain the appeal of herbicides as a labour-saving technology (Moseley & Pessereau 2022).

At the household level, the women farmers we interviewed for our study kept highlighting how difficult it was for them to find enough labour to maintain their own farm fields (Moseley & Pessereau 2022). There are at least three reasons that help us understand why women are facing labour constraints. The first was just discussed above and has to do with the overall labour limitations facing agriculture that have been exacerbated by the exodus of young people to work in the artisanal gold mining sector. The second has to do with labour dynamics

---

7. In fact, control of gold mining contributed to the success of the Ghana, Mali and Songhay Empires that flourished between the nineth and fifteenth centuries (discussed in Chapter 4).

within households. At the household level in rural Burkina Faso, women have limited abilities to requisition labour for work on their own fields because work on the large household fields managed by the senior male head of household is prioritized. Women may be able to get their own children to work on their farm fields in the early mornings and evenings (outside of work on household fields that occurs in the middle of the day), but their chances of getting other household members to help them is limited.

In order to understand household gender dynamics, some background is useful. In the rural areas of Burkina Faso, most farming households are relatively large, male-headed, extended families. These households include the eldest male and his wives, his adult male children and their wives, and their children. This is a polygynous[8] society where wealthier and older men may have up to four wives if the household is Muslim or animist (although this is forbidden for Christians). Roles and responsibilities are highly gendered[9] and, within the household food economy, women are responsible for cooking (often done in rotation with other women in the household) and for providing the sauce ingredients[10] for meals, whereas men provide the staple grains. This means women farm and that they tend to grow or forage a lot of sauce crops (peanuts, cowpeas, okra and hibiscus in farm fields; vegetables in gardens; foraged tree crops like baobab leaves and African locust tree pods). Women also cultivate rice for weddings, funerals and sale.

The third and final factor that may be exacerbating labour constraints at the household level is more environmental. Weeding is a very labour-intensive process and the women we interviewed suggested that the weed problem is much worse than it used to be. Weeds do tend to be more of a problem on soils that have lost their fertility and women tend to be allocated plots that are more "tired" by men in the family (who control the distribution of land). Herbicide resistant weeds are also a documented and growing problem in Burkina Faso (Heap 2014).

The women we interviewed indicated that they had increasingly turned to herbicides since 2011 or 2012 to solve their labour problems. They suggested that herbicides were now much more affordable and easier to find. Interestingly, the application of herbicides is highly gendered (like other forms of labour) and tends to be performed by men. As such, most women would get a male

---

8. Some scholars argue that polygyny in the African context (the practice of a male having multiple wives) emerged as a result of depopulation from slavery and warfare (Manning 1981).

9. As a theoretical framework, feminist policy ecology acknowledges that gender is a social construction but argues that socially constructed gender roles have real material consequences for the ways that people interact with the environment and other people.

10. Sauce is the nutrient dense gravy that accompanies a starch in a traditional meal. Examples of traditional sauces include peanut sauce, baobab leaf sauce and *soumbala* (fermented locust tree seeds) (Morgan & Moseley 2020).

relative to apply the herbicides with a backpack sprayer or hire a man to do the job (although there were some instances, 11 per cent of interviewees, where women applied herbicides themselves when other options were not possible). While increasing herbicide use is a rational response to labour constraints, it also contributes to growing health risks as well as the spread of herbicide resistant weeds. As discussed earlier, many of the herbicides are glyphosate-based (a suspected carcinogen and hormone disruptor) or atrazine-based (a known teratogen and hormone disruptor). Most women we interviewed were not aware of health concerns related to herbicide use, although they often indicated that they would not apply them when young children were present (suggesting some level of concern). Ten per cent of women in our surveys did cite health risks and concerns, whether for themselves or for infants they nurse or carry on their backs throughout the day. Some women noted how herbicide backpack sprayers often leak onto people's backs, and they were afraid that their contact with the chemicals would get their infants sick.

Conclusion

In sum, a New Green Revolution for Africa (GR4A) rice project in southwestern Burkina Faso, even though it was focused on food crop production for domestic markets (and not an inedible export crop like cotton), did not improve the nutrition of participating women farmers. While arguably good for the Burkinabè agronomists and seed companies involved, such top down, productionist efforts have missed the mark on addressing food insecurity and malnutrition and arguably have the potential to make it worse. Furthermore, a history of cotton cropping in the same area planted the seeds for a conflagration of herbicide use, fanned by cheap prices, a loose regulatory environment, an exodus of young rural workers, and women's limited ability to command household labour. In Chapter 9, I will explore how thinking about food security in terms of broader food systems and rural food environments in Burkina Faso offers reasons for hope.

6

PROBLEMS WITH THE RICARDIAN FOOD SECURITY DREAM: BOTSWANA'S CONUNDRUM OF GROWTH WITH HUNGER

Abstract ideas and theories are powerful and some have had a profound influence on the way African economies are organized. David Ricardo (1772–1823) was a British economist who was an ardent supporter of free trade. More specifically, he developed the concept of comparative advantage, the idea that countries should specialize in those industries where they have the greatest efficiency of production relative to alternative uses of their resources. In a hypothetical example, he famously showed how England and Portugal would both benefit from a trading relationship if England specialized in cloth production and Portugal in wine production, even though Portugal could produce both more cheaply (Ricardo 1817). Some 160 years later, Ricardo's ideas about free trade and comparative advantage would serve as the backbone of the neoliberal economic policies that were foisted on the continent as part of the World Bank and the IMF's structural adjustment packages. More remarkably perhaps, some of these ideas also influenced the organization of African economies that were not indebted and forced to take on such reforms, such as Botswana.

Botswana is considered to be an African development success story and is imagined by some to be one example of an ideal future for the continent. Under this "Ricardian food security dream scenario",[1] well run governments judiciously manage export-based economies emphasizing commodities for which they have a comparative advantage. In other countries, agriculture adopts the full suite of New Green Revolution technologies and large commercial farms produce the majority of food for urban populations. The problem is that resource-based economies are often undiversified and produce deep inequalities. Furthermore, high-input agriculture, while good at producing lots of food, is also heavily dependent on fossil fuel energy, water for irrigation, and is largely performed by wealthy farmers. As energy prices rise, so does the cost of food production. The

---

1. Thanks to the late Rachel Shurman (University of Minnesota) for suggesting the "Ricardian food security dream" portion of this title in a personal conversation.

combination of expensive food and deepening inequality means that hunger will persist even if Africa's leaders do everything "right".

Botswana now faces such a conundrum. It has a well-developed export sector for gem quality diamonds and also earns significant income from high-end eco-tourism and beef exports. Despite its wealth and good governance, the structure of Botswana's economy has contributed to growing inequality, gender disparities and food insecurity. Furthermore, given its high dependence on imported food (90 per cent), Botswana finds itself particularly vulnerable to fluctuating global food prices. Women farmers, in particular, have been neglected by the export-oriented livestock industry and crop farming has become less of a priority since the country moved away from a food self-sufficiency paradigm in the 1980s. The development of boreholes for the male-oriented cattle industry has also often left women farmers struggling to find adequate water resources (Moseley 2016; Fehr & Moseley 2019). This chapter explores the history and development of this conundrum shaped to some extent by the ghost of Ricardo.

Understanding Botswana

At independence in 1966, Botswana was arguably one of the least likely African countries to become an economic success story. At that time, it was suffering from drought, the country had less than five kilometres of paved roads and it had only 23 college graduates. This backwater British protectorate known as Bechuanaland did not even have a capital city as it was administered from a town in South Africa called Mafikeng (now known as Mahikeng). Furthermore, as discussed in Chapter 2, the British thought of Bechuanaland as little more than a labour reserve to supply workers to the mines in South Africa. Today Botswana is a middle-income country that is considered to be a model of economic development. The name Botswana in the Setswana language literally means the country of the Tswana. While the Tswana are the largest ethnic group (73 per cent), other major groups include the Kalanga (18 per cent) and Basarwa (2 per cent) (Parsons 2023).

Among other factors, geographer Abdi Samatar (1999) argues that social cohesion among Tswana elites, and a sense of duty to the nation, explains some of the country's success. There are important examples of this agency in the colonial and postcolonial periods. In the nineteenth century, the mining magnate Cecil Rhodes famously wanted to absorb then Bechuanaland under the administration of the Cape Colony for which he served as prime minister. Four Tswana chiefs were opposed to this given Rhodes' history in the region. As a result, three of the chiefs (with the support of missionaries) departed for England in 1895 in hopes of meeting with the prime minister to express their opposition to Rhodes' annexation plans. Not gaining an audience with the prime minister, the trio went on a speaking tour to garner support from the British public and were eventually able to get a meeting with Queen Victoria. While historians debate the impact of the visit,[2] Bechuanaland did remain a separate protectorate under direct British rule rather than being absorbed into Rhodes's empire. This effort and agency of the three chiefs is memorialized today via a statue in Gaborone known as the Three Dikgosi Monument (Parsons 1998).

Botswana gained independence in 1966 and its first president was Seretse Khama, a descendent of one of the three chiefs previously discussed. Khama was known for his shrewd management of the economy, diplomatic caution vis à vis powerful neighbours and careful use of resources: all important as his country was dirt poor when he ascended to the presidency. One poignant example of his frugality occurred when he travelled to a meeting of the Organization of African Unity (OAU) in Addis Ababa early in his presidency. All throughout the day, African presidents had been flying into Addis in their private jets when the organizers learned that Khama would be arriving on a commercial flight. They quickly cordoned off first class and escorted the passengers from the plane onto a red carpet and across the runway to the arrivals terminal (assuming Khama would be among them). Meanwhile Khama, who had been in economy, had walked over to passport control and was waiting in line like the rest of the crowd. General alarm ensued once his presence was discerned by passport control, and they made him walk back to the plane and then down the red carpet (Samatar 1999). More important than his frugality were important decisions Khama made to champion good governance, investment in education and development, all of which were made more possible by the discovery of diamonds a few years after independence, aspects of which will be discussed in the next section.

Two important and interconnected decisions that Khama made early on were that the government must be run by competent individuals who were appointed based on merit rather than family connections, and that investments in education were essential for training a skilled workforce. With regards to the former, Khama made the unusual decision to hire expatriates to fill some government positions early on, including ministerial posts, until local people could be trained to take on these roles (Hope 1995). Establishing a university in Botswana was also a major priority. One of my favourite stories concerning

---

2. The impact of the visit is debated because the failed Jameson raid happened around the same time. Rhodes, then prime minister of the Cape Colony, instigated the botched Jameson raid into the Afrikaner-controlled Transvaal from Rhodesia in late 1895 in hopes of triggering an uprising of British workers. The uprising never happened and it turned British public opinion against Rhodes who was removed as prime minister of the Cape Colony. This turn of opinion, coupled with the visit of the three Tswana chiefs to England, also scuttled Rhodes' plans to annex Bechuanaland (Rotberg 1988).

the establishment of the University of Botswana (where I taught as a visiting professor in 2012) has to do with the original fund-raising effort to bankroll the school. Khama asked every family in the country to donate a cow (with wealthy families giving more) to help establish a university in the country. The herd that was assembled was then driven down to South Africa for auction and the proceeds were used to establish the university. A statue now stands outside the University of Botswana library (see Figure 6.1) commemorating this act of generosity and belief in the value of education. Today, Botswana's proportion of government expenditures on education rank among the highest in the world, the literacy rate is 87 per cent and there is universal free primary education (World Bank 2023a). A legacy of Khama's competent governance strategy, combined with an approach that built on local democratic traditions (Samatar 1999), is that Botswana has had a consistent track record of free and fair elections since independence. Botswana's fifth and current president is Mokgweetsi Eric Masisi (World Bank 2023a).

In terms of its geography, Botswana is a semi-arid country covering 581,730 km² (22nd largest in Africa) and is one the least densely populated in

Figure 6.1 Statue on the University of the Botswana (UB) campus commemorating the original cattle drive that funded the establishment of the university
Source: Photo by author.

the world with 2.7 million people. Its landscapes are dominated by the Okavango Delta in the north, the Kalahari Desert in the southwest, and grassland savannas in the southeast (where the majority of the population lives) (see map of the country in Figure 6.2). Botswana's agriculture is overshadowed by the livestock sector. While most food is imported, the main food crops grown in the country are maize, sorghum, millet and pulses (mainly cowpeas, Bambara groundnuts and mung beans). Interestingly, watermelon may have originally been domesticated in Botswana, likely in the Kalahari area by the Basarwa people (University

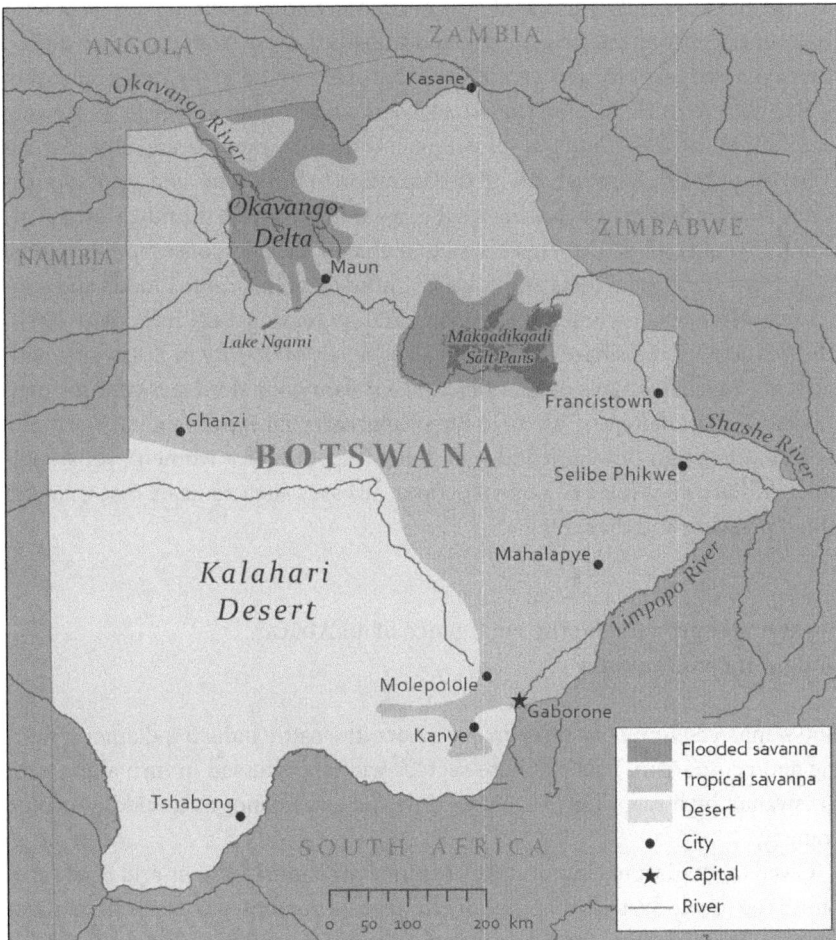

Figure 6.2 Ecological zones of Botswana
Cartography by Julia Castellano, Macalester College.
Sources: World Wildlife Fund 2012; Stanford University 2013; Simplemaps 2023; Esri Africa 2018; World Bank 2018; World Agroforestry Centre 2014; UTM Zone 34S.

of Missouri 2020). The livestock sector is dominated by men, although women are more active with small stock such as chickens, and crop farming is overwhelmingly female. The traditional rural configuration in Botswana is spatially composed of three components: the village (the centre of social activity; the lands (outlying areas for crop cultivation); and cattle posts (areas further out for herding activity) (Bryant *et al.* 1978).

Botswana's traditionally rural population has rapidly transformed over the past several decades and was reported at just over 72 per cent urban in 2022, up from 4.4 per cent at independence in 1966, and 16.5 per cent in 1980 (World Bank 2023d). This rapid rise in urbanization tracks very closely with the country's economic growth rate and transition away from crop agriculture. Botswana's largest city and capital is Gaborone with an estimated population of 232,000 in 2023. As Botswana had no capital at independence, Gaborone was built in the 1960s and is a sprawling, low density, car-oriented city (Gwebu 2003; Marr 2019). Associated with urbanization in Botswana, and across southern Africa for that matter, has been a dietary shift known as the nutrition transition (Nnyepi *et al.* 2015). This refers to a change in food consumption habits, away from traditional diets and towards more meat, processed foods and sugars that often accompanies urban living and increasing wealth (Popkin 2004). These changing diets have led to increasing levels of obesity in Botswana: with current prevalence rates of 32.1 per cent for women and 9.7 per cent for men (Global Nutrition Report 2023). In Botswana, carrying extra weight is culturally desirable as a sign of wealth and well-being (especially for women), yet associated with growing rates of non-infectious diseases such as heart disease, high blood pressure and diabetes.

Spectacular growth and the emergence of an African middle-income country

Botswana's economy has three major pillars: the cattle industry, diamond mining and ecotourism. Each of these sectors will be discussed in turn along with Botswana's high rate of economic growth and ascendance as a middle-income country.

Given the traditional importance of animal husbandry in semi-arid Botswana, one of the first industries that the government set about developing was the livestock sector (Samatar 1999). In order to control foot and mouth disease (which is easily spread from wildlife to cattle), the government oversaw a large fencing campaign. They also introduced strict veterinary controls and the construction of government run abattoirs for the butchering of meat (all under the coordination of the Botswana Meat Commission). Other innovations included the drilling of bore holes for the watering of cattle (a process that had started in the colonial period) and the conversion of commonly held pasture land into private property in order to encourage investment (Peters 1984). All of these changes led to tremendous growth in the livestock sector and the development of an export-oriented beef sector. The export of beef to the EU was facilitated by the agreement that gave Botswana preferential access to EU markets (Darkoh & Mbaiwa 2002).

While livestock production and beef exports grew tremendously, many of the aforementioned changes had some notable social and environmental consequences. In Botswana's dryland environment, livestock and wildlife historically moved across large areas in search of grazing areas. Fencing prohibited the seasonal movement of wildlife and led to overgrazing in some cases (Darkoh & Mbaiwa 2002). The drilling of boreholes also led to the congregation of cattle in certain areas, resulting in degradation, and overtaxed aquifers in some situations. The conversion of grazing commons to private property was a local land grab of sorts and a huge windfall for wealthy elites who tend to own more cattle (Peters 1994). Lastly, given the fact that agricultural activities in Botswana are highly gendered, and cattle rearing is viewed as a male occupation, men were the main beneficiaries of government investment in the livestock sector (Hovorka 2012).

The other big development in Botswana was the discovery of diamonds just a few years after independence. President Khama quickly realized that his country did not have the expertise to extract and manage this new-found resource and that he would have to deal with the behemoth next door, the DeBeers diamond cartel in South Africa. While DeBeers had a track record of taking the upper hand in negotiations with new players in the diamond business, Botswana shrewdly hired a former DeBeers employee to help them negotiate a 50/50 split of revenues from the operation, a deal that was unprecedented at the time (Samatar 1999). Over time, Botswana would become the leading exporter of gem-quality diamonds in the world, and it would use these resources to expand government services, including education and healthcare. It was also able to avoid the resource curse that has bedevilled many other African countries that have hit the natural resource jackpot.

The resource curse is the idea that countries with valuable resources succumb to corruption, mismanagement, the decline of democratic governance and a simplification of the economy (Collier 2007), although critical scholars are careful to point out that it is politics and political economy that drive this problem rather than the resource itself (Moseley 2009; Watts 2009). Countries like Nigeria and Angola, both major oil exporters, have been cited as prime examples of this phenomenon (Collier 2007). While Botswana is often held up as an example of a country that has avoided this problem (Samatar 1999), and it has managed to mostly avoid egregious corruption, there are concerns that the economy has been unable to sufficiently diversify beyond its dependence on diamond extraction (Hillbom 2008, 2011). Diamonds do account for 90 per cent of Botswana's exports, and it has struggled to create a more diversified economy (World Bank 2023a). This is not to say that Botswana has not attempted to develop new industries. Realizing that it could earn more, diversify the economy and create new high paying jobs if it exported cut rather than raw diamonds, in recent years Botswana has tried to develop its own diamond-cutting industry. This has been no easy task as much of the world's diamond cutting expertise is concentrated in Belgium and India, but Botswana has spent considerable sums to bring in foreign cutters and train up its own workforce. Botswana now holds back 15 per cent of its raw diamond exports for domestic cutting (Moseley 2015b).

The last main pillar of the Botswana economy is ecotourism, which is mostly centred around the Okavango Delta in the northern part of the country and nearby Chobe National Park, which is known for its large and significant elephant population (see Figure 6.2) (Moseley 2013b). Before developing its tourism sector, Botswana made a strategic decision that it would develop a high-end, low-volume approach. As such, it decided to eschew the backpacker tourist trade that was common in neighbouring countries and focus on those tourists who were willing to pay top dollar for a luxury safari experience. The advantage of this strategy was that it mitigated the environmental impacts of higher-volume approaches, while still ensuring that the country generated significant revenue from tourism. In fact, many well-heeled tourists fly into the northern provincial town of Maun (adjacent to the Okavango Delta) and never see the capital city Gaborone or the southeastern part of the country where most Motswana[3] live (see Figure 6.2) (Mbaiwa 2017a).

All thee pillars of the Botswana economy have contributed to the country's significant growth rate over the past 50 years, with the diamond sector being the most significant driver. Between 1975 and 2022, the country's economy (measured as GDP per capita) grew at an average annual rate of 3.42 per cent (World Bank 2023b). This growth rate is on par with that of the Newly Industrialized Countries (the NICs) of the Pacific Rim, yet we almost never hear about Botswana. As seen in Figure 6.3, Botswana's growth in GDP per capita also far exceeded that of the average in Sub-Saharan Africa (SSA). Today Botswana is a middle-income country with foreign exchange reserves of over five billion USD (equal to 8.2 months of imports as of March 2023) (CEIC Data 2023a). For comparison, the United Kingdom's foreign

---

3. Motswana is the local term for people from Botswana.

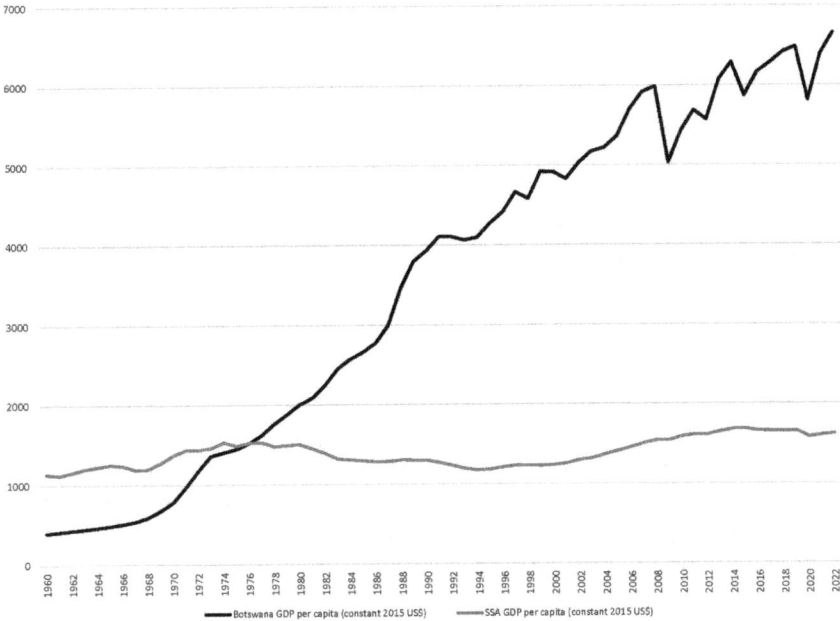

Figure 6.3 Botswana's economic growth relative to other African countries (1960–2022)
Source: Chart created using data from World Bank (2023b).

exchange reserves equated to 1.7 months of imports in that same time period (CEIC Data 2023b). In many ways, the Botswana story is a remarkable one, and it deserves to be celebrated, especially at a time when prejudice often colours how African countries and peoples are perceived. In fact, I often hold up Botswana in my own teaching as a shining example of success, highlighting the agency and smart decisions of the country's leaders. Nonetheless, even economic success stories need to be studied carefully because policy-makers seeking to emulate them must be wary of potential pitfalls. The reality is that there is a dark side to Botswana's amazing story, and it has to do with inequality and hunger.

Growth with hunger

What Botswana's tremendous growth rate masks is persistent inequality and a remarkable level of poverty and food insecurity for a middle-income country. Below I examine Botswana's income inequality and its structural drivers, changing agricultural development approaches, food insecurity and the relationship between all three of these factors.

The structural drivers of inequality

One common measure of income inequality is the Gini coefficient.[4] This measure ranges between 0 and 100 with higher scores suggesting greater inequality. Botswana's most recent score is 53.3, ranking it tenth in the world behind other countries in the region like South Africa at 63 (#1) and Namibia at 59.1 (#2). The global average is 38 (World Bank 2023c). Perhaps more interesting is why Botswana, a poor country at independence, grew to have such high levels of inequality in the postcolonial period. I would argue, and so have others, that the structural conditions in the country, or the nature of Botswana's economy, offer at least a partial explanation for this situation (Moseley 2012b; Besada & O'Bright 2018). In other words, consistent with the political ecology framework introduced in the first section of this book, the material conditions on the ground are related to broader political economy. All three of Botswana's major economic pillars, the cattle industry, diamond mining and high-end ecotourism, have fostered a distorted income distribution, which I examine below.

As a semi-arid country, with a traditional agropastoral economy, cattle are culturally important and a store of wealth for the country's elite. As such, while it probably made some sense for the country to focus on this traditional sector of the economy as a development strategy, investments in the livestock sector (ranging from boreholes, to veterinary infrastructure, to the marketing of Botswana beef) overwhelmingly benefitted wealthier families and men who own the most cattle (Good 1993; Darkoh & Mbaiwa 2002). These initiatives not only increased the revenue streams of wealthy Motswana but helped them consolidate their power and expand their land holdings when commons were privatized and fences erected (in the name of veterinary controls) (Peters 1994). Even though agriculture represents a smaller part of Botswana's economy today, many of Botswana's urban-based elite continue to own cattle and regularly visit their cattle posts on the weekend.

The diamond industry in Botswana is well run with several open pit mines that meet the highest international social and environmental standards (Torres Solís & Moroka 2011). The reality, however, is that diamond mining in Botswana is highly mechanized and employs a relatively small number of people, limiting the ability of this industry to better distribute wealth through employment creation. As compared to the cattle industry, the wealth generated by the diamond mining is arguably

---

4. The Gini coefficient is based on the Lorenz curve. A Lorenz curve basically plots the total income in a country (on the vertical axis) against proportions of the population (on the horizontal axis). A 45-degree line would be perfectly distributed income with, for example, 50 per cent of the population controlling 50 per cent of the income. A more bowed out line shows more unequally distributed wealth with say 50 per cent of the population only controlling 20 per cent of the income.

more evenly distributed as it funds a significant share of government operations, including social welfare payments, healthcare services and education, from which a large swath of the population benefits. Nonetheless, because elites tend to be better educated, they capture most of the higher paying, public sector jobs that compose a big part of Botswana's economy (Mogalakwe & Nyamnjoh 2017).

Lastly, the ecotourism industry also helped concentrate wealth into the hands of a few. Many of the safari-companies and lodges in Botswana are foreign-owned, which means that many of the profits are repatriated outside of the country. Furthermore, those tourism businesses that are owned by Botswana nationals, tend to be naturalized, white Botswana citizens (many of whom came from neighbouring Zimbabwe and South Africa) who are already in the upper income strata. Lastly, unlike diamond mining, tourism has created a fair amount of employment, but a large share of these positions are lower-paying service sector jobs, such as room cleaners and restaurant workers (Mbaiwa 2017b).

What income inequality has resulted in is a persistent level of poverty in Botswana. According to the UNDP (2023), 17.2 per cent of the population in Botswana suffered from multidimensional poverty (a combination of health, education and standard of living measures)[5] in 2021, while another 19.7 per cent was classified as vulnerable to multidimensional poverty. Narrower, income-based measures of poverty suggest that 15.4 per cent of the population lives below the international poverty line which is $2.15 per day (2017 PPP), and 38 per cent below the lower middle-income country poverty line which is $3.65 per day (2017 PPP) (World Bank 2023e).

Botswana's pivot away from food self-sufficiency and its increasingly vulnerable food system

In the early 1980s, during the administration of Botswana's second president Quett Masire, Botswana made the decision to pull back on its support for crop agriculture based on the advice of its neoliberal economic advisors.[6] These advisors argued that it made more sense for Botswana to focus on its diamond exports and to abandon the goal of food self-sufficiency or the objective of producing as much food within

---

5. According to the UNDP (2023), they measure multidimensional poverty by examining overlapping deprivations for ten indicators across three dimensions: health, education and standard of living. There are two indicators each for the health and education dimensions and six indicators for standard of living.

6. As Botswana did not become indebted in the late 1970s like many other African countries; it was not forced to take out loans and adopt structural adjustment policies. Nonetheless, its leadership did adopt some neoliberal economic policies while pushing back on others.

its borders as possible (Lado 2001). This advice for Botswana to focus on what it did well, and to trade for those goods that it produced less efficiently, was consistent with Ricardo's ideas about comparative advantage discussed in the introduction to this chapter. While Botswana was never completely self-sufficient in terms of food production, not surprising for a largely semi-arid country, it went from producing about two-thirds of its own food at independence to importing around 84 per cent of its cereal grains (mostly wheat and maize), and 90 per cent of its food overall, in the 2018–22 period on average (GIEWS 2022; FAO 2023). This decision to pull back from crop agriculture disproportionately impacted women and the rural poor, the main smallholder farmers (see Figure 6.4). While crop agriculture had never been a significant economic engine for the country, it was a food security safety net for many of the rural poor. This decision highlights how many policy-makers and advisors think about agriculture as a tool for economic growth rather than a strategy to support food security for more marginalized segments of the population.

The other interesting, agriculture-related transition over the past 30 years has been a shift from drought tolerant small grains like millet and sorghum to maize. At first blush, this appears to make little sense as maize is more vulnerable to drought, a situation that is occurring with increasing frequency in Botswana due to climate change. However, our interviews with smallholder farmers (who were largely older females) revealed that labour constraints explained much of the change as maize was easier to care for than millet and sorghum (Moseley 2016).[7] This shift was also facilitated by the greater levels of research investment that have gone into developing maize varieties. The upshot of both these shifts, disinvestment in crop farming and the rise of maize, is a more anaemic crop farming sector in Botswana that is highly vulnerable to the vagaries of climate change.

Food insecurity

The combination of income inequality and persistent poverty on the one hand, combined with declining and more variable local food production and a dependence on food imports on the other, means that poor Motswana are highly vulnerable to food insecurity when there are global food price shocks and the price of imported food rises. As was discussed in Chapter 3, food security has six dimensions (availability, access, utilization, stability, sustainability and agency), and several of those dimensions have been undermined by the structural conditions impacting Botswana's food system. Because of persistent poverty, the

---

7. Sorghum and millet are more susceptible to predation by birds than maize. Keeping the birds away from the former requires a small army of children who are now less available to help women farmers in rural areas.

Figure 6.4 A Motswana woman on her smallholder farm in Botswana
Source: Photo by author.

ability of low-income, urban Motswana to access or purchase food is limited, and this is made more challenging when food prices rise. Furthermore, while rural households used to produce much of their own food, this is now less and less the case because of disinvestment in agriculture, and these households are also buying a larger share of their nutrition. It is also the case that the sustainability of local food production has been undermined by a transition to maize, a crop that is more vulnerable to rainfall fluctuations.[8] Lastly, the stability of local food prices and supplies is vulnerable to global fluctuations because of a heavy dependence on food imports. This dependence on food imports has not gone unnoticed and is a topic of conversation in Botswana's local media (e.g. Hore 2022).

While considerable scholarly and policy-maker attention was paid to the 2007–08 global food crisis (with a 50 per cent increase in prices), there is somewhat less scholarship on the food price increases in 2010–12 that impacted a number

---

8. Maize cross pollinates with other maize plants to produce the seeds or kernels that humans consume. This cross-pollination typically occurs in a two to three week window during the summer which becomes problematic if it coincides with drought conditions.

of African countries, including Botswana.[9] In February 2011, the FAO reported that their food price index had risen 6.5 per cent from the previous year, with the index reaching a level not seen in 20 years (FAO 2012). Some analysts have connected these food price rises to the Arab Spring (Rosenberg 2011) and the famine in the Horn of Africa in 2012 (Moseley 2012c). In light of these price rises, in 2012 I launched a small study of urban, peri-urban and rural households in southeastern Botswana (where the majority of Motswana live) to better understand their food security in a country that had become highly dependent on food imports. I administered surveys to: 89 urban households in three lower income neighbourhoods in the capital city Gaborone, 30 households in a peri-urban on the outskirts of Gaborone and to 39 rural households in two villages in the southeastern part of Botswana. These randomly selected households were stratified by wealth and asked a series of questions about their income and expenditures, proportion of income spent on food, level of food insecurity (using the HFIAS scale[10]), the extent to which high food prices are a problem, coping strategies to deal with food shortfalls, and (in the case of urban interviewees) potential impacts of their coping strategies on extended family members living in rural areas.

My findings from the 2012 research on food insecurity in Botswana were not encouraging given the country's level of development and status as a middle-income country. As seen in Table 6.1, urban respondents spent 35 per cent of their income on food, peri-urban respondents 26 per cent and rural respondents 61 per cent. These are quite high proportions to be spending on food and suggested limited manoeuvrability were food prices to climb higher. For context, the average UK resident spent about 16 per cent of their household budget on food in 2023 (NimbleFins 2023). For better or for worse, one of the most common ways that urban residences were decreasing expenses in light of higher food expenses was to reduce remittances to their relatives in rural areas (Moseley 2016). This was problematic because it meant that rural areas, which already tended to be poorer than urban areas, might have borne the brunt of the 2010–12 food price increases.

In my surveys, the levels of food insecurity (as measured by HFIAS) were considered to be moderate in all three geographies (urban, peri-urban and rural),

---

9. The food price increases of 2010–12 likely received less attention than those in 2007–08 for a couple of reasons. First, the increases in 2007/2008 came after a relatively long period of price stability, so it was more of a shock. Second, while the absolute peak was higher in 2011, the level or rate of increase was lower. The 2010–12 price increases were less novel, having come just a few years after the 2007–08 increases.

10. We also used this measure as part of our food security studies in Burkina Faso discussed in the previous chapter. The household food insecurity access scale (HFIAS) is based on the answers to nine qualitative questions about the previous four weeks. These range from more moderate questions (how many days did you worry about not having enough food et eat) to more severe (how many times in the past 4 weeks did you go 24 hours without eating). The score ranges from 0 to 4, with higher scores indicating greater levels of food insecurity (Coates et al. 2007).

Table 6.1 2012 food security survey findings for Botswana

Sample size	Demographic	Mean per capita income per day		% income spent on food	Food security	
		Pula	USD		HFIA score	Qualitative
89	All Urban	25.5	$3.55	35%	2.8	Moderate
30	Urban lower tercile	1.71	$.24	38%	2.9	Moderate
30	Urban middle tercile	16.9	$2.35	35%	2.9	Moderate
29	Urban upper tercile	59.14	$8.21	31%	2.5	Moderate
30	Peri-urban	40.1	$5.56	26%	2.8	Moderate
39	Rural	12	$1.67	61%	3.2	Moderate

Note: The household food insecurity access score (HFIA) is tabulated based on the scores to nine questions in: 1 = food secure, 2= mild food insecurity, 3 = moderate food insecurity and 4 = severe food insecurity.

Source: Author surveys.

but considerably worse in rural areas. This was supported by additional findings (in Table 6.2) showing that 76 per cent of rural, 67 per cent of peri-urban and 55 per cent of urban respondents were facing moderate to severe food insecurity. The food secure population varied from 18 per cent in rural areas, to 24 per cent in urban areas, to 27 per cent in peri-urban areas. In sum, while 2012 was a bad year in terms of global food prices and this was clearly impacting the poor in Botswana, this level of food insecurity in that year was disconcerting given that Botswana is supposed to be a development success story. It also underscores how vulnerable the country was because of what I would call its Ricardian food security approach or a trade-based approach for maintaining food supplies.

Unfortunately, in the intervening ten years since I did these surveys, the situation has not greatly improved. Ndhleve et al. (2021) undertook food security surveys in Botswana in 2021 using the HFIAS measure. While they did not break down their findings by rural, peri-urban and urban areas, they found 21.3 per cent to be food secure in the country as a whole (Ndhleve et al. 2021). While different than the HFIAS, the 2022 Global Food Security Index (0–100 with higher scores being better) initiative gave Botswana a score of 51.1, this compares to 61.7 for South Africa, 51.9 for Mali and 49.6 for Burkina Faso (Economist 2022).[11] That Botswana would rank behind Mali is a bit shocking given the vast wealth

---

11. According to the report's methodology section (Economist 2022), the Global Food Security Index (GFSI) is based on food affordability, availability, quality and safety, and sustainability and adaptation. It measures these different dimensions via 68 indicators. These GFSI dimensions intersect to some degree with the six dimensions of food security presented in Chapter 3, leaving out agency and utilization.

Table 6.2 Proportion of Botswana survey households facing food insecurity in 2012

Sample size	Demographic	Food secure	Mild food insecurity	Moderate food insecurity	Severe food insecurity
89	Urban	24%	20%	12%	44%
30	Peri-urban	27%	7%	23%	43%
39	Rural	18%	6%	19%	57%

Source: Author surveys.

differences. To be fair, the government of Botswana does offer robust safety nets for some segments of their population (especially compared to other African countries). The main form of support in rural areas is drought relief when an official dry year has been declared. In urban areas, there is a public works programme, known as *Ipelegeng* in Setswana, that employs those who fall below the poverty line and are able bodied (Moseley 2016).

Conclusion

Even though Botswana is considered to be an economic success story that is known for its good governance, and it has astutely managed its diamond resources, the country is plagued by unacceptably high levels of food insecurity (especially for a middle-income country). The main pillars of Botswana's economy, the cattle industry, diamond mining and ecotourism, have fostered inequality and persistent poverty among segments of the population. This poverty has clearly compromised some household's ability to access or purchase sufficient amounts of healthy food. Furthermore, this situation was made worse by the government's decision to abandon food self-sufficiency as a policy goal. While this may have been reasonable from a neoliberal economic standpoint, and Ricardo's concept of comparative advantage, it made less sense if one considered agriculture's household food security and poverty alleviation benefits for more marginalized segments of the population. Given Botswana's heavy reliance on food imports, local food prices often rise and fall in relation to increasingly volatile global food price trends, putting Botswana's poor at risk of food insecurity. Greater levels of household food production would at least buffer them from these shocks. Experiments with backyard gardening, and support of women's agriculture, in Botswana will be explored as a reason for hope in Chapter 10.

7

BIG AGRICULTURE'S TAKEOVER
OF SOUTH AFRICA'S LAND
REDISTRIBUTION PROGRAMME

While doing research in the Western Cape Province of South Africa on land reform in 2005, I sat down with the head of the commercial farmers' association, or Agri South Africa (AgriSA), for an interview. He was a white farmer and agronomist by training whose family had lived in South Africa for generations. Towards the end of a long discussion on the dynamics of commercial farming in the region, he became a bit philosophical. He believed that large-scale commercial farmers in the region (practically all white at that time) had the knowledge and expertise to turn the region into a major breadbasket, or to use his own words, "to make the region bloom". Undergirding this statement was an almost evangelical belief in the strengths of conventional, industrial agriculture as it had come to be practised in much of the Global North. As if his words were prophesy, it was around this time that I would also begin to learn about white, Afrikaans-speaking South African farmers moving into other African countries like Mozambique and Zambia to set up new commercial farming operations. By 2010, AgriSA would be engaged in negotiations for long-term land leases or land acquisitions in 22 African countries (Hall 2012).

Closer to home, these same ideas about the superiority of the large-scale commercial farming model would also reshape South Africa's approach to land reform. In post-apartheid South Africa, one of the most significant initiatives of the new government in 1994 was land reform. While the programme initiated under Nelson Mandela (known as Settlement/Land Acquisition Grants (SLAG)) had its problems, it was somewhat flexible about how land could be used by recipients and had a strong focus on poverty alleviation. Under Thabo Mbeki, South Africa's second president who took office in 1999, the land reform programme was restructured and rebranded as Land Reform for Agricultural Development (LRAD), becoming more commercially oriented. These changes reflected the discursive power of the commercial agriculture community in South Africa that framed their approach to farming in terms of modernity and progress – and alternatives as backward and primitive. This chapter provides background on South Africa, its intertwined colonial and agricultural histories, as well as details on its land restitution and redistribution efforts.

Understanding South Africa

Unlike the food security and agricultural development dynamics of other African countries examined in this book, South Africa was different because of its history of white settler colonialism. Not only have European settlers been in the country for over 370 years, but they still represent a significant minority population (about 7.7 per cent in 2022, 18.1 per cent in 1980 (Stats SA 2022)), and controlled the government up until 1994. As such, the dynamic between African and European farmers, and African and European farming knowledge, is arguably longer and more complicated in South Africa than in other African countries.

With the exception of a few areas in the northeastern most part of South Africa, the vast majority of South Africa falls within the temperate zone, that is south of the Topic of Capricorn at 23.5°S. There is a fairly strong correlation between biomes (areas of broadly similar climate and vegetation) found in South Africa and the types of agriculture practised (see Figure 7.1). There are five major areas. First, a Mediterranean zone in the southwestern part of the country with wet winters and dry summers. This is a horticultural area (fruit and vegetables) and the country's major wine producing region. Second, there are semi-desert areas (similar to the Sahel in West Africa) in the central southwest of the country where animal husbandry and mixed farming is practised. Third, temperate grasslands in the central part of the country south of Pretoria and Johannesburg where large-scale, mechanized wheat and maize farming has been dominant. Fourth, a more tropical region on the southeastern coast (influenced by the warm Indian Ocean) where large sugar cane plantations and some smallholder farming is found. Fifth, in the Northeast and along the southeastern coast there are semi-forested savannas where lots of small holder agriculture and animal husbandry traditionally has been practised, with maize as the major food crop. The first four areas have historically been controlled by white farmers whereas the fifth area and some of the fourth have had more black smallholder farmers. These last two areas are also where the communal areas, or former homelands, are largely located (Meadows 2000).

The Dutch East India Company established Cape Town in 1652 as a waypoint and refreshment station between Europe and its trading and colonial interests further east (e.g. Indonesia, Sri Lanka). There they encountered the Khoisan people, two distinct yet related populations that were engaged in hunting and gathering (the San) and cattle herding (the Khoikhoi). While Bantu peoples inhabited other areas of South Africa in much denser numbers, and would likely

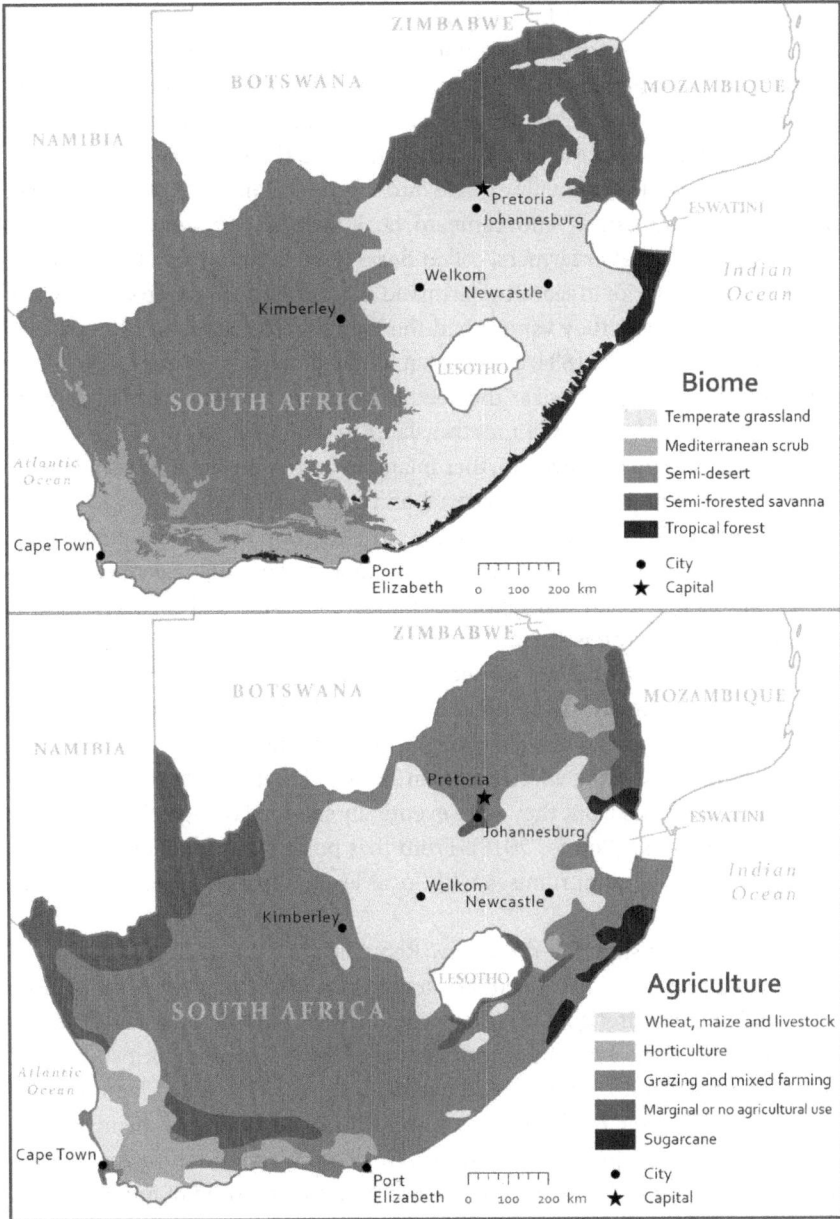

Figure 7.1 Climate and agricultural zones in South Africa
Cartography by Julia Castellano, Macalester College.
Sources: World Wildlife Fund 2012; University of Texas 1979; Simplemaps 2023; Esri Africa 2018; World Agroforestry Centre 2014; UTM Zone 34S

DECOLONIZING AFRICAN AGRICULTURE

have crushed the Dutch at first contact, the more thinly dispersed Khoisan put up less resistance and were more vulnerable to European diseases (such as a deadly small pox outbreak in 1713). In fact, the sad reality is that many Khoisan fell victim to the genocidal tendencies of Dutch settlers who frequently hunted down and killed the Khoisan with no consequences (Laband 2020).

Cape Town would eventually expand into a settler colony with Dutch, German and French immigrants[1] who came to be known as *Afrikaners* (Dutch for Africans). The Afrikaner farmers, called *Boers* (Dutch for farmer), progressively moved into the interior in search of farmland. The British took control of the Cape Colony in 1795 (after they vanquished the Dutch in the Battle of Muizenberg) and abolished slavery in 1834, a move that irritated many Afrikaners. Due to the limitations on slavery, as well as the need to secure more fertile farmland, many Boers, aka *voertrekkers* (sometimes translated as pioneers), left the Cape Colony around 1836 and pushed even further inland to areas outside of British control, establishing the Republics of Transvaal and the Orange Free State. Along the way they encountered African resistance. The Khoisan and Xhosa people they encountered on the periphery of the Cape Colony were not of sufficient density to put up significant resistance beyond cattle raiding and scattered attacks (Lester 2000). As they pushed into the highveld (an upland area) and further east into Natal, the voertrekkers encountered denser populations of Zulu speakers (Zulu speakers were also widespread in the Natal Colony administered by the British in the southeast). The Zulu Kingdom put up highly organized and effective resistance against the voertrekkers in 1838 and then in subsequent engagements with the British, but they were eventually subdued by superior fire power (guns and cannons) (Lohnes 2018). From that point forward, colonial officials usually worked out compromises with local leaders to administer these areas (Lester 2000).

Initially the British ignored the new Republics of Transvaal (also known as the Republic of South Africa) and the Orange Free State, but then became more interested in them after gold was discovered in the Transvaal near what is today Johannesburg in the late nineteenth century (1886). Growing British interest in the Transvaal and its gold would eventually lead to the failed Jameson raid discussed in Chapter 6 (a failure that contributed to Botswana being able to maintain its separate protectorate status), two Boer Wars, and the incorporation of these areas into British-controlled South Africa (composed of four areas: the Cape Colony, Natal, the Orange Free State and Transvaal).

---

1. Many of the French immigrants were Huguenots, or French protestants, who were forced to leave France by the Catholic monarchy. It is this group that played a significant role in establishing the wine industry in the southwestern Cape.

Labor struggles and racial capitalism

The demand for labour in the gold mines, and more importantly in the diamond mines near Kimberly[2] controlled by Cecil Rhodes, would eventually create huge labour problems for the white farmers in the Cape Colony. Since the abolishment of slavery, white farmers had struggled to maintain enough labour on their farms. Unwilling or unable to pay higher wages, a strategy these white farmers developed to maintain a grip on their labour supply was the *tot* or *dop* (Afrikaans for an alcoholic drink) system. This involved giving farm workers a ration of dreg wine (the wine the farmers could not sell) several times a day. The resulting chemical dependency that developed over time made farm workers less likely to leave, especially when combined with other strategies to foster debt bondage, such as selling goods to farm workers on credit at the farm store (DuToit 1994). When the mining sector boomed in the late nineteenth century, and demand for workers created higher wages, the farmers responded by doubling the *tot* to the equivalent of two bottles of wine per day per worker (Scully 1992).

The other challenge white farmers faced in the nineteenth century was competition from black farmers, especially after the abolition of slavery. Colin Bundy's (1979) groundbreaking research revealed that black farmers, either through work on their own farms or via share cropping arrangements, were often better and more efficient farmers than white farmers. Furthermore, many black Africans refused to work on white farms, especially in good rainfall years when they were producing plenty of their own food (Bundy 1979). This fact, that white farmers were struggling to compete with black farmers, and could not acquire sufficient labour, eventually led to the 1913 Native Land Act (passed when the British were still in control[3]), a structural solution to the white farmers' competition and labour problems. The 1913 Native Land Act restricted black South Africans access to land to a few designated areas known as native reserves, also referred to as homelands or *Bantustans*. The homelands were sufficiently small (13 per

---

2. Diamonds were discovered in Kimberly in the 1860s in an area known as the Griqualands. The Griqua were a group of mixed-race people (of Khoisan, European and other ancestry) who lived on the periphery of the Cape Colony and practised animal husbandry and extensive farming. This area was incorporated into the Cape Colony by the British after Diamonds were discovered. Cecil Rhodes started out as a small-time miner, but then was able to gradually buy out the other miners, eventually controlling 90 per cent of the mining (Waldman 2006, 2007).

3. The fact that the 1913 Native Land Act was passed when the British were in control (South Africa became a sovereign state within the Commonwealth in 1934) is an important detail because it highlights how the British were complicit in South Africa's racialized regime. Some commentators associate Apartheid with the Afrikaner led National Party that came to power in 1948 (a group that did double down on policies to separate racial groups and usher in an era known as high Apartheid), implicitly absolving the British of their role in Apartheid.

cent of the land area for over 80 per cent of the population) that it was impossible for black South Africans to support themselves through farming. This structural situation forced black South Africans to migrate from their homelands to work off-farm (in order to survive) on white-run farms or in the city for low wages.

The situation created by the 1913 Native Land Act was the epitome of an arrangement known as racial capitalism. The concept of racial capitalism is often associated with the pioneering work of Cedric Robinson (2000). Robinson questioned Karl Marx's assertion that capitalism was a new form of labour relations that represented a revolutionary break from feudalism (Marx 1926). Rather Robinson argued that labourers had long been depicted as racial subjects (even in medieval Europe) in order to extract labour from them at the lowest possible rate or terms (Robinson 2000; Salvato 2022). As Ruth Wilson Gilmore (2020) has articulated: "capitalism requires inequality, and racism enshrines it". South Africa's 1913 Land Act was a critical foundation for constructing a more exploitative system of racial capitalism. In rural areas, what unfolded was also an extreme version of what is known as agricultural dualism. Agricultural dualism exists when there is a small-scale, subsistence-oriented farming sector that exists in parallel with large-scale, commercially oriented agriculture. Under this arrangement, which is not uncommon in the Global South, large-scale farmers exploit the labour of small-scale farmers knowing that they can pay them less than a living wage because of their subsistence activities (Moseley *et al.* 2013).

The quiet neoliberal coup

The conservative, Afrikaner-led National Party (NP) came to power in 1948, ushering in the era of high Apartheid, or the most extreme version of segregationist policies. The African National Congress (ANC), created in 1912, predates the NP and struggled for years against the racist Apartheid system. By the late 1980s, it was becoming increasingly clear that the Apartheid system was imploding on itself and not sustainable as a result of ANC resistance, international sanctions, a bloated bureaucracy and the sheer fact that it was extremely expensive to maintain artificially constructed social divides. It is at this time that then NP leader F. W. DeKlerk started quietly negotiating with the imprisoned ANC leader Nelson Mandela. Mandela was released from prison in 1990 and triumphantly led the ANC to victory in South Africa's first free and fair elections in 1994 (Mandela 1995).

The ANC, and Nelson Mandela himself, had long taken an interest in Marxist interpretations and policy prescriptions. When the United States under Ronald Reagan was backing P. W. Botha's South African Apartheid regime in the 1980s, in the name of anticommunism during the Cold War, Cuba under Fidel Castro was supporting the ANC. Furthermore, central to the ANC's political platform was the 1955 Freedom Charter (a statement of core principles of the ANC) which stated that: "the land should be shared amongst those who work it" (Britannica 2023).

The ANC's 1994 land charter included a pledge to acknowledge and address the history of stolen land in South Africa.[4] Given the ANC's Marxist leaning history, several policy shifts in and around the time the ANC came to power may come as a surprise. According to geographer Dick Peet (2002a,2002b), a powerful Academic-Institutional-Media (AIM) complex emerged in this period that wielded considerable discursive power. During the transition, business leaders, academic economists and representatives from international financial institutions (IFIs) quietly and repeatedly met with Mandela's ANC leadership team. In these meetings, they presented dire economic scenarios, white and capital flight, if the ANC were to engage in land expropriation or the nationalization of certain parts of the economy. One of their prime examples of what could go wrong was the case of neighbouring Zimbabwe where the economy had increasingly faltered, and whites and their money had poured out of the country, as Robert Mugabe pushed fast track land reform (which involved the seizure of white farmsteads) in a desperate attempt to maintain popular support and cling to power (McCusker *et al.* 2016b). These representatives also cleverly presented the National Party (NP) as an interventionist regime (and it is true that the NP provided significant support to white farmers, one of their main political constituencies). Using the policies of Zimbabwe and the NP as foils, the AIM representatives effectively lobbied for a more free market and non-interventionist approach to agriculture and land reform (not to mention in other segments of the economy).

Land reform

The upshot of the ANC's last-minute, neoliberal turn was that the government adopted a market friendly approach to land reform based on the World Bank's

---

4. Language from the 1994 Land Charter: "We the marginalised people of South Africa, who are landless and land hungry, declare our need for all the world to know. We are the people who have borne the brunt of apartheid, of forced removals from our homes, of poverty in rural areas, of oppression on the farms and of starvation in the bantustans. We have suffered from migrant labour that has caused our family life to collapse. We have starved because of unemployment and low wages. We have seen our children stunted because of little food, no water and no sanitation. We have seen our land dry up and blow away in the wind, because we have been forced into smaller and smaller places. These are the biggest difficulties facing our country in the future". Cited from the Land Charter adopted during the Community Land Conference in 1994.

negotiated land reform model involving the principle of willing seller/willing buyer. Under this approach, the government would work with the market to redistribute land, providing grants to members of historically disadvantaged groups to buy land that white farmers willingly put on the market (Moseley & McCusker 2008). There were at least three problems with this model that will be discussed further below: (1) it was incredibly expensive and therefore incremental and slow; (2) it did not redistribute wealth as white farmers were being compensated in full for their land; and (3) the farms that were sold tended to be the most marginal that could not make it in the newly competitive agricultural sector. For agriculture more broadly, the ANC chose to embrace the global market place and pull back on agricultural subsidies (a reaction to NP policies that were more interventionist). On the one hand, this shift was great for some farmers and farm sectors, such as the South African wine industry that grew by leaps and bounds after the end of Apartheid and the shelving of international sanctions against South African products (Moseley 2008c). On the other hand, large numbers of South African farms went out of business in this time period, leading to a great deal of consolidation in the agricultural sector (Moseley 2007b). This newly competitive marketplace also made it challenging for new farmers to emerge.

South Africa's land reform programme had three major components: land restitution; tenure reform and land redistribution. The land restitution programme was for those who could legally establish that they had lost their land following the 1913 Native Land Act. While many have filed to get their land back, the vast majority of cases have been settled via cash payments rather than actual restitution. The other problem is that many families are unable to legally document the loss of their land (even when it clearly happened) and/or they lost their land long before 1913. Recall that the Dutch had been in Cape Town since 1652, and European settlers had been stealing land from black South Africans long before the 1913 Native Land Act. The tenure reform programme was largely focused on the former homelands or what came to be known as the communal areas in the post-Apartheid era. Here the focus was on establishing clear usufruct or use rights to the land (discussed in Chapter 1), not to be confused with land privatization efforts that have happened in other African countries.

The third and final land reform programme, and arguably the most significant, was land redistribution. Under South Africa's first majority-rule president, Nelson Mandela (1994–99), the programme offered Settlement/Land Acquisition Grants (SLAG) to historically disadvantaged groups. These grants were for 16,000 rand, or around $4,000 at the time, to be used to purchase land (Moseley 2007b). There were positive and negative aspects to this programme. The positive dimension (depending on your perspective) was that the programme had a poverty alleviation focus and was quite flexible in terms of how beneficiaries could use the land after it was purchased. Many urban, black South

Africans were not interested in farming but rather acquired land in order to establish a small homestead and/or pursue other economic activities. The critiques of the programme were numerous. First, the grants were small enough that several households often needed to pool their grants together in order to purchase property, leading to the challenges of joint ownership. Second, because there was only one grant per household, male heads of household tended to be the main beneficiaries and women lost out. Third, some saw it as a problem when the land acquired was not used for agricultural purposes. Fourth, there was no technical support or additional funds for those who wanted to develop a new farming operation. Lastly, the programme was considered to be moving at too slow a pace. By 1999, they had redistributed 1 per cent of white-owned farmland, when the goal was 30 per cent by that year (DuToit 1994). In general, this first phase of land redistribution was seen as inadequate, if not a failure.

The commercial turn in South Africa's land reform efforts

When Thabo Mbeki came into office, South Africa's second president (1999–2008), the land redistribution programme was paused in order to address some of the problems and concerns with SLAG discussed above. What emerged after this reflection period was a restructured and rebranded programme known as Land Reform for Agricultural Development (LRAD). The new programme allowed individuals rather than households to apply for grants, which addressed the concern about male bias with SLAG. The grants were also bigger, so fewer households would need to pool grants in order to purchase a farm, and offered on a sliding scale from 20,000 to 100,000 rand. The grants were allowed to increase in size in relation to a beneficiary contribution of cash, existing assets or labour (aka sweat equity). Finally, LRAD recipients could also apply for additional funding from the Commercial Agriculture Support Program (CASP) from the Department of Agriculture for training and inputs.

Arguably one of the most significant changes reflected in the LRAD programme was its new commercial orientation. In contrast to SLAG, LRAD grant applications needed to include a business plan for how the beneficiaries would run the farm with a clear commercial orientation. Other types of agriculture, such as smaller scale, non-commercially oriented farming or non-agriculture related uses, were not considered acceptable and would be denied. Also lost in this transition was the poverty alleviation focus of SLAG in favour of agricultural development. Here the idea was to use land reform to train a new generation of commercially oriented black farmers. These changes reflected the discursive power of the commercial agriculture community in South Africa, as represented by the commercial farmers association (AgriSA) highlighted in the introduction

(Hall 2004, 2010). AgriSA had clear ideas about land reform, articulated in their publications and voiced to national, provincial and local government. In 2000, AgriSA wrote: "AgriSA supports land reform in principle ... We do however feel strongly that agricultural land should as far as possible be retained for agricultural use and production. We are not in favour of residential type developments on farmland. Agricultural land should be farmed in an economically viable and environmentally sustainable manner" (AgriSA 2000: 2).

They framed their approach to farming in terms of modernity and progress, and implied in their message was the idea that African farming approaches were backward and primitive. In AgriSA's framing, returning black South Africans to small-scale agriculture was a "liberal white fantasy", whereas they sought to help black farmers build wealth through commercial farming (Hall 2010). Today the perceived superiority of large-scale, industrial farming is essentially a hegemonic discourse[5] in South Africa. This narrative occludes the fact (discussed previously) that the system was propped up by the exploitation of black labour and could not compete with indigenous African farming expertise absent an unlevel playing field.

Struggling land redistribution

In 2005, I began a new research project in the Western Cape Province of South Africa examining farm workers engagement with the LRAD land redistribution programme that had recently been initiated under President Thabo Mbeki. Farm workers, many whom had lived for generations on white-run farms in the Western Cape, were the main intended beneficiaries of the unfolding LRAD initiative in the region. The research project had two main objectives: to document the farming knowledge of the farm worker population; and to assess the results of the LRAD programme five years into its implementation.

Unlike other areas of South Africa, the Western Cape Province has no homelands. This is related to the province's long history of European encroachment and genocidal tendencies towards the local Khoisan population. Today the remnants of the Khoisan population persist in a mixed-race population (often referred to as "coloureds" in South African parlance). Many of the farms in the Western Cape continue to be white-run and the majority of the farm workers are of this mixed race group. The majority of farm workers historically lived on the farms where they were employed. However, while now some farm workers still

---

5. The idea of a hegemonic discourse was briefly discussed in Chapter 3. Hegemonic discourses are essentially dominant explanations that have been broadly accepted, even among those who are painted in a negative light by such discourses.

live on-farm, many have moved off-farm and into nearby communities. This is because some well-intentioned farm worker protections put in place after the fall of Apartheid perversely led many white farm owners to let go of permanent farm workers and rehire them as temporary workers in order to avoid regulations (Du Toit & Ewert 2002). Nonetheless, in terms of spatial layout, the traditional commercial farm in the Western Cape might have a large farm house where the white farmer lived, and then the farm workers quarters tucked back out of sight over the hill (see Figure 7.2).

The first part of this research project sought to document the knowledge of farm workers still employed on commercial farms in the Western Cape Province. To do this, we interviewed 60 farm workers and 20 white farm owners working on a range of different types of commercial farms: dairy and livestock, fruit, vegetable, wheat and wine (see Figure 7.3). A persistent perception in post-Apartheid South Africa was that farm workers were mindless, unskilled labour. Furthermore, this same narrative suggested that it was the prowess of white farmers that had made the commercial farming sector successful (Du Toit 2004). This narrative is not unique to South Africa but exists in other regions of the world where settler colonialism abounds. Judy Carney (2002)'s groundbreaking text *Black Rice*, for example, revealed the role of enslaved Africans in bringing rice farming expertise and knowhow to the Americas (previously portrayed as skilled white farmers directing black labour). Furthermore, other

Figure 7.2 Traditional Western Cape farm
Source: Photo by author.

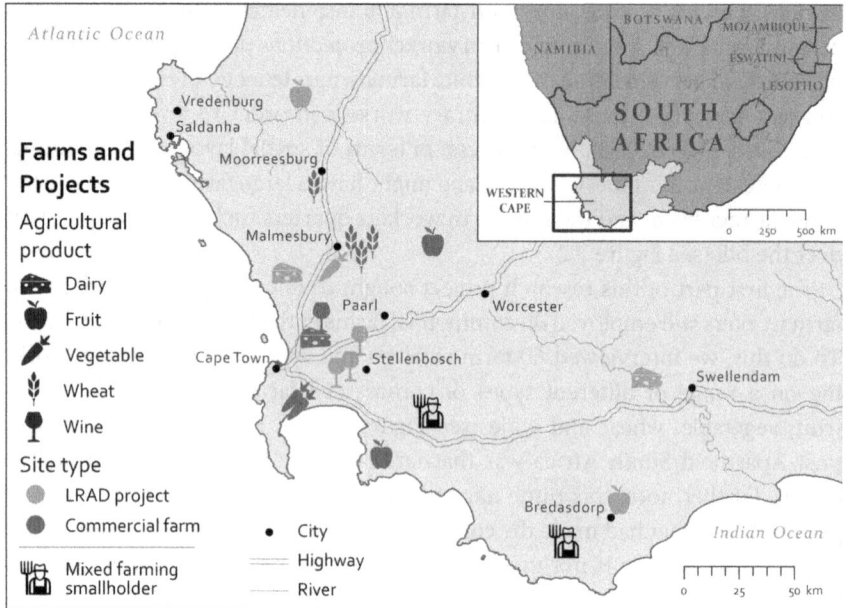

Figure 7.3 Location of farms and land redistribution projects in the Western Cape (where interviews were conducted by author)
Cartography by Julia Castellano, Macalester College.
Sources: Humanitarian Data Exchange 2020; SEDAC 2010; Simplemaps 2023; Esri Africa 2018; World Agroforestry Centre 2014; UTM Zone 34S.

emerging research in South Africa at the time suggested that "white farming narratives" had greatly understated African agency in the management of commercial farming landscapes (Du Toit 1993) and that the farming practices of white settler colonialists often degraded farming soils and rangelands (Archer 2000). Unfortunately, subaltern knowledge is often buried or repressed due to centuries of oppression. Our interviews revealed that farm workers knew much more than they were given credit for and possessed highly specialized and technical skills. A good example of the latter was the ability to quickly and proficiently prune fruit trees and grape vines. Inexperienced or careless workers could damage vines or trees, influencing production, a problem that was becoming more common as white farmers transitioned away from a permanent workforce and used more contract labourers. Many farm workers also knew about the interactions between pests, crops and chemical usage, including problems such as pesticide resistance. Lastly, the dynamics of soil erosion were well understood as well as the practices needed to manage this problem (Moseley 2006a, 2007b).

After documenting farm worker knowledge, we also sought to understand how farm workers, who had benefitted from LRAD land redistribution grants, were faring as new farm owners. At the time of the study, there were 101 active LRAD projects in the Western Cape. We chose to focus on those projects that were at least three years old so there was a track record to examine, which immediately cut the number of projects down to 37, from which we randomly selected 16 for in-depth examination. Of these farms, some were wholly owned by the farm workers who had pooled their grants to buy them, whereas others were jointly owned by the farm workers and the original farm owner (known as share equity schemes). Under such shared ownership arrangements, the farm workers typically owned the majority of the business (e.g. 70 per cent) and the white farmer the minority portion. The thinking behind joint ownership was that the original farm owner with business experience would share his[6] knowledge with the emerging farmers. As one might imagine, this was often a complicated relationship because there was a tendency to fall back into the old roles of owner and worker. Many LRAD projects also involved commercial bank loans, whether the projects were share equity schemes or fully owned by the beneficiaries. These loans were necessary because it was rare that the LRAD grants could cover the full cost of the farm as farmland was extremely expensive in the Western Cape. Such loans also added pressure for LRAD farms to be sufficiently profitable that they could pay back loans in a timely fashion.

An analysis of project documents, several visits to the farm (see Figure 7.4), and interviews with at least five beneficiaries led us to qualitatively classify the 16 farms in terms of five categories: (1) sold back (2 farms); (2) not operational (2 farms); (3) questionable (2 farms); (4) some promise (5 farms); and (5) more promise (5 farms).[7] With about 40 per cent of the LRAD projects failing or near failing after three years, and 60 per cent succeeding, the record of land redistribution efforts in the Western Cape was clearly mixed. Very disconcerting was the significant minority of farms that had either stopped functioning or been

---

6. While women obviously play huge roles on farms, all of the white farm owners or co-owners in our study were men, which is the general pattern in South African rural, white society.

7. "To be classified as 'sold back,' a project not only needed to have been sold, but the author needed to visit the farm and speak with at least five of the (former) project beneficiaries. In some instances, the author was able to determine that a farm had been sold, but was unable to visit the farm or make contact with a sufficient number of beneficiaries. [...] To be classified as 'not operational,' the farm was found to still be in the hands of the beneficiaries, yet there was no farming activity occurring on the land. A classification of 'questionable' was made if the farm operation was clearly losing money, there appeared to be a decline in farming activity, but the farm was still operating. An LRAD farm was found to have 'some promise' if there was farming activity occurring and the financial situation was tenuous, yet still at a point where the farm could become economically viable. The last and best category was 'more promise.' In such instances, there was farm activity occurring, the farm was generating enough revenue to meet recurring financial obligations and there was potential for growth and profits" (Moseley 2007b: 8–9).

Figure 7.4 Land reform beneficiaries running a small dairy near Swellendam, South Africa
Source: Photo by author.

sold back (the latter often being foreclosure as a result of failure to make loan repayments).

More broadly, I concluded that the reason so many LRAD farms were failing had to do with one or more of the following factors, which could be categorized as either structural in nature (related to underlying conditions, political ecology and political agronomy) or more technical. There were at least three underlying structural conditions that mitigated against programme success. First, given the ANC's embrace of neoliberal economic policies and globalization, the commercial farming sector had become incredibly competitive. This sector was already undergoing a great deal of consolidation (the number of commercial farming units declined by 21 per cent between 1993 and 2001) and now new and emerging farmers were being asked to participate in this hyper-competitive environment with no protections, even though white farmers had historically benefitted from considerable government support (Cousins 2006). Second, and as discussed earlier in this chapter, the World Bank model of market-based land reform, based on the principle of willing seller/willing buyer, meant that more marginal farms (in terms of productive potential) were more likely to be sold to new black farmers as these were the white-owned farms facing financial pressure in this newly competitive environment. As such, it was not surprising that land reform beneficiaries would also struggle to make these farms financially viable (made all the more challenging if commercial bank loans were used to cover part of the cost of the farm). Third, while the LRAD grants were larger than those offered by the previous programme, they were still sufficiently small that large numbers of beneficiaries needed to pool their grants in order to buy a farm. While some groups were able to work well together, others were not. Furthermore, as a commercial enterprise, it was near impossible that a farm could support such a large number of people, often numbering from 50 to 100. Obviously, the grants could have been larger, but this also raises questions about paying full market value for farms. This not only saddles new farmers with debt, but has made land redistribution a slow and expensive process (Moseley 2007b).

In addition to structural issues, there were also some more technical challenges. First, sometimes there was a mismatch between the nature of the farm and the skills of the workers. For example, the workers might have been very good at growing vegetables, but those talents were less useful on a dairy farm. Furthermore, given the commercial orientation of the LRAD grant application process, applicants were often encouraged to develop a business plan based on the original orientation of the farm rather than their own skill set. Second, given the history of Apartheid, and the exclusion of people of colour from the business side of managing farms, many farm workers and other land redistribution beneficiaries lacked the business experience and training needed to run a successful commercial farm. Third, those LRAD beneficiaries involved in share equity schemes sometimes faced problems with the white farmer co-owner, yet there was very little legal recourse to resolve such problems. For example, one of the LRAD projects we studied involved a white farmer who refused to share management of the farm with the new worker co-owners, ran the farm out of business, and then it was repossessed by the bank. The workers in this case had no legal recourse for pushing the white farmer to share management information and decision-making (Moseley 2007b). Fourth, new farmers, especially those that produce fresh fruit and vegetables, were often challenged to establish deals with local supermarket chains that preferred consistent supplies at volume. This problem has grown over time as the grocery retail business has become more concentrated in South Africa, creating problems for farmers and consumers (Peyton *et al.* 2015). Lastly, the slow pace of bureaucracy at the Department of Land Affairs and the Department of Agriculture often meant that new projects did not receive their funding for critical inputs (e.g. fertilizer) in a timely fashion, which means they missed critical time windows that reduced crop production (Moseley 2006a, 2007b).

Plus ça change?

LRAD was replaced in 2006 by the Proactive Land Acquisition Strategy (PLAS). The main differences from LRAD were that the government purchased farms on the open market and then allocated them on leasehold tenure to beneficiaries (usually a 30-year rental arrangement) (Rusenga 2022). This gave the state more control over the purchase of farmland, rather than distributing grants to beneficiaries who purchase the land, and eliminated some of the problems of beneficiaries becoming indebted and losing farms. The downside was that the state has used this arrangement to more strictly require new farmers to adopt a largescale commercial farming model. Regrettably, the idea of subdividing large farms into smaller plots was still anathema to policy-makers, suggesting that the power of the agribusiness lobby remained stronger than ever. The other problem was that new farmers' tenure was also more tenuous under the new arrangement as beneficiary farmers no longer owned the land (Hall & Kepe 2017). The longstanding goal for land reform in South Africa has been to redistribute 30 per cent of white owned farmland to historically disadvantaged groups. The original deadline for this target was 1999, and then that was pushed back to 2014 and now 2030. As of 2021, it was estimated that 17 per cent of land had been redistributed (Sihlobo & Kirsten 2021). The slow pace of reform is largely due to the market-based land redistribution model which is costly and inherently incremental as transfers may only occur when land is put on the market.

The other idea that has repeatedly been discussed in South African policy circles is expropriation without compensation (EWC) or the taking of land without compensation (but more likely below market compensation in practice) (Fiamingo 2021). Proponents see this as a way to speed up the land reform process, whereas critics continue to hold up Zimbabwe as an example of how this tactic will inflict irreparable harm on the economy. In 2017, at the ANC policy conference, a proposal was made to amend the constitution to allow for expropriation without compensation under a narrow set of circumstances (ANC 2017). This same idea was also reiterated in a 2019 report by the Presidential Advisory Panel on Land Reform and Agriculture (RSA 2019). An "expropriation bill" was passed by the South African parliament in September 2022 (Merten 2022) and the debate has continued up through August 2023 on the pros and cons of amending the constitution to allow for this. AgriSA, the commercial farmers association, remains steadfastly opposed to any such change, highlighting unencroachable private property as central to successful commercial agriculture (Daily Investor 2023).

Conclusion

Land reform in South Africa has long been seen as critical for historical redress for black South Africans who lost their land and livelihoods during centuries of white settler colonialism. Despite land reform's central place in the ANC's political platform, it was essentially hijacked by neoliberal thinking in the 1990s and then redesigned in the early 2000s as a conventional agricultural development programme. The commercial farmers association's influence on this programme, and on South Africans' thinking about agriculture more broadly, has been omnipresent over the past 30 years. The political agronomy framework, presented in Chapter 3 of this book, helps us understand the discursive power of big agriculture in this context. Furthermore, political ecology both highlights the discursive hegemonic power of mainstream agriculture, as well as the processes that marginalized both black Africans and their knowledge, and the situated agency of those pushing for change. This last theme, opportunities for change, will be explored in Chapter 11 where I examine new models of smallholder-based land reform and agrarian justice.

PART III

REIMAGINING AFRICAN FOOD SYSTEMS

8

FARMER AND CONSUMER AGENCY IN MALI: FOOD SOVEREIGNTY AND AGROECOLOGY AMIDST GLOBAL FOOD PRICE FLUCTUATIONS AND CONFLICT

I spent the summer of 2009 interviewing women who prepared food for urban households in Mali's capital city, Bamako. The years 2007 and 2008 had been rough ones for many of West Africa's lower income, urban households as global food prices had gone through the roof, especially for rice whose price had risen 100 per cent. I wanted to learn how these households had dealt with this food crisis and why food riots had not broken out in Bamako as they had in many other West African cities. In the midst of this research, one warm afternoon I remember sitting under a tree in a family compound in a lower income neighbourhood of Bamako talking to the women of a household. I was going through my usual interview questions asking them about their food choices and strategies for dealing with higher prices. While broken rice imported from Asia had become the staple across urban West Africa (see Chapter 5 on Burkina Faso), these women were insistent that they preferred a locally produced rice variety known as *Gambiaka* (hailing from the Office du Niger area discussed in Chapter 4). I drilled down a bit on this choice because I knew *Gambiaka* rice was often more expensive. Why, I asked, were they willing to pay more for this local rice? At that point, an older grandmother on the periphery of the conversation piped up and explained, in the most animated fashion, "because the imported rice doesn't make me feel well and makes my stomach hurt".

This preference for local rice, and this grandmother's negative experience with imported rice, likely reflects two factors. First, there is a longstanding rice consumption tradition in Mali as this is where African rice was originally domesticated (Carney 2002). Second, the broken rice imported into West Africa is often of inferior quality. In fact, some of this rice has been found to be 30–40 years old and of minimal nutritional value, having been dumped onto markets after national governments turnover and sell their old security or buffer grain stocks (Tondel *et al.* 2020). As such, the food preferences of our grandmother make total sense in this context. The women in this interview also went on to explain that they would purchase locally plentiful sorghum if they could not afford local rice, avoiding the problem of expensive imported rice.

These acts of food consumption agency in urban Mali (a key dimension of food security discussed in Chapter 3) operate in parallel with the agency and grit of Mali's smallholder farmers. In many areas of Mali (and West Africa more broadly), smallholder farmers have never quite trusted the state. This mistrust is at least partially due to the legacy of colonialism and its attempts to break local food production strategies in favour of more commercially oriented agriculture. As such, Mali's farmers have often kept one leg in the subsistence world as a way to feed their families and manage risk, and another leg in commercial agriculture in order to earn some money. While some agricultural development experts have framed subsistence production as a problem,[1] and dismissively characterized Malian farmers as risk averse, this is really a case of farmers exercising their agency by quietly asserting their rights to produce their own food in ways that make the most sense to them, and pushing back on unfair commercial arrangements.

The 2007–08 period was momentous for Mali's farmers and food consumers. At the start of 2007, Malian activists, farmers, and international peasant organizations met in southern Mali to develop the Nyéléni Declaration, an important statement of principles regarding food sovereignty. In that same period, and as discussed above, the global food crisis would hit urban households across West Africa. Fortunately for Mali's consumers, their farmers had decided to grow a lot more sorghum that year due to their unhappiness with global cotton prices. Furthermore, Mali also continued to produce much of its own rice and was not as dependent on imports. Unlike some neighbouring countries, Mali's urban consumers preferred local rice if they could afford it (like our grandmother in the introduction) or pivoted to sorghum as the cost of imported rice sky rocketed. In both cases, the food sovereignty movement, and Malian urban consumer behaviour, social movements and agency were a key counterbalancing force to globalization and hegemonic discourses regarding development and modern agriculture. Furthermore, these social movements were not just pushing back in a material sense (e.g. opting for locally produced grains or refusing to grow cotton at low prices), but they were producing powerful counter narratives involving new paradigms such as food sovereignty.

This chapter explores Mali's somewhat unplanned experiment with food sovereignty that came about as a result of increasingly well-organized smallholder farmers and urban consumers who preferred local grains. The food crop of choice for poorer households, sorghum, provided a number of agroecological

---

1. Subsistence production has frequently been framed as a problem by advocates for a New Green Revolution in Africa (Page 2012).

benefits: it was more amenable to intercropping, it required few purchased inputs and it was more drought tolerant. As Mali's security situation has deteriorated, and the ability of the central government to provide services has declined, this has also raised new questions about what makes for a more robust food system in areas directly and indirectly impacted by conflict.

The Nyéléni Declaration

As discussed in Chapter 3, more than 500 food system representatives gathered in the village of Nyéléni in southern Mali in February of 2007. They came from around the world and from organizations representing peasants, family farmers, fisher folk, indigenous peoples, landless peoples, rural workers, migrants, pastoralists, forest communities, women, youth, consumers and environmental and urban movements. In the midst of another global food crisis fostered by growing trade in the food crops and the loss of farmer agency, they gathered to assert the principles of the peasant-led and increasingly global food sovereignty movement. From this meeting emerged the Nyéléni Declaration in which they defined food sovereignty as "the right of peoples to healthy and culturally appropriate food produced through ecologically sound and sustainable methods, and their right to define their own food and agriculture system" (Sélingué 2007).

This international meeting occurred in the small, rural village of Nyéléni in southern Mali for a couple of reasons. The first was symbolic. Historically, the village is named after a locally famous and legendary women of the same name. Nyéléni means first daughter in the local language, Bamanankan. She was the only child of a rural couple who became highly regarded as a farmer, supported her family and is credited with the development of a local grain *fonio* (which is sometimes referred to pejoratively in English as "hungry man's rice"). According to oral history, Nyéléni won a traditional agricultural weeding contest associated with the *ciwara*, the mythical antelope that taught the Bamanan ethnic group how to farm.[2] By winning the contest (which entails quickly and cleanly weeding a portion of a field with a hand-held hoe or *daba*), and defeating her male competitors in the process, Nyéléni became even more revered as a farmer (Grain 2007). As such, the symbolism of this name was particularly important for the many women farmers present at this meeting (Nyéléni Declaration 2007: 13). Furthermore, social movements, which are an important component of food

---

2. The *ciwara* is celebrated in traditional dance and in the aforementioned weeding contests (both of which I was able to observe in the 1980s (Moseley 1993). Given that the *ciwara* was associated with traditional farming and animist religious practices, it has been marginalized by more mechanized farming practices and Islam.

sovereignty, agroecology and the agency dimension of food security, need strong narratives (such as the history of Nyéléni) to counter the hegemony of the conventional agricultural development paradigm.

The second reason for holding this meeting in Mali was more contemporary and related to the increasingly strong peasant farmers' movement in Mali and West Africa more broadly. Succumbing to pressure from farmers' organizations, the government of Mali had recently agreed to make food sovereignty a policy priority. According to an interview with Mamadou Goita, a local food sovereignty activist, "[w]e realised that the giant food corporations were taking advantage of the WTO negotiations on trade in food, and of all the talk about food aid, to gain control over food production worldwide and to make everyone dependent on them for food" (Grain 2007: 13). In that same interview, Goita went on to say:

> [W]ho are we producing for? Are we producing export crops? This is what is happening in most countries in West Africa. Farmers are producing cash crops to have money in their pockets and no one cares about producing food for the local population. Take Benin, Burkina Faso, even Chad. In these countries the best-organised crop is cotton. The decision-makers are not putting money into staple foods such as maize, sorghum and millet. (Grain 2007: 14)

At the regional level, Mali's smallholder farmers were also being supported by the Network of West African Peasant and Agricultural Producers' Organizations (or ROPPA) (McKeon *et al.* 2004; McKeon 2020). As such, given the food sovereignty movement's momentum in Mali at that time, this made the country an attractive location for such an international meeting.

The global food crisis of 2007–08

In 2007–08, many urban areas in West Africa were dealing with the impacts of the global food crisis, described previously as a time period when the average price of food rose 50 per cent and rice prices rose 100 per cent. The latter was particularly significant as rice had had become a key staple in many West African cities, including those in Mali. Unlike the situation in Burkina Faso, where rice is produced less (discussed in Chapter 5), the position of rice in Mali's food system varies geographically, being a staple in some areas and not in others. African rice varieties had originally been domesticated in Mali's inner Niger delta (Carney 2002), had been cultivated in the river's food plains for centuries and was a staple for river plain populations. In contrast, in the southern parts of the country, where Bamako is located, rice was traditionally more of a luxury food reserved for weddings, funerals and religious holidays, with sorghum, millet and maize serving as the main staple grains. But urban Mali had undergone some significant dietary transitions, with rice becoming a major urban staple over the previous 30 years. Many food scholars might describe this shift as consistent with the nutrition transition, the idea that as populations become wealthier and more urbanized, they start to consume more meat, fat, sugar and processed foods (Popkin 2004).

While the nutrition transition model is helpful to some extent for understanding changing urban diets in Mali and neighbouring countries, I believe a political ecology framework provides more insight as it describes the shifting structural conditions, as well as household level actions, that led to the prominence of rice in West African urban diets, including those in Bamako. What we learned after extensive research in 2009 (Moseley *et al.* 2010; Moseley 2011) is that important structural changes at the international, regional and national levels led to growing levels of Asian rice flowing to West African ports, and significant shifts at the household level fostered increased rice consumption. At the international level, the first wave of the Green Revolution led to the rise of new Asian rice exporters, namely Thailand and Vietnam. While this same initiative also bolstered rice production in India and China, much of that rice was destined for their large internal markets (Moseley 2013a). At the regional and national levels, the neoliberal economic reforms known as structural adjustment, discussed in Chapters 2 and 4, led to a reduction in tariff barriers in Mali and surrounding countries that made it cheaper to import rice from south and southeast Asia. Finally, many of the subsidies that had been in place for rice farmers were reduced because of neoliberal economic reforms, depressing domestic rice production in West Africa. Interestingly, however, in Mali supports for rice farmers in the Office du Niger schemes were never reduced as much as they were in neighbouring countries like Côte d'Ivoire and The Gambia (Moseley *et al.* 2010). This meant that Mali continued to produce more of its own rice than its neighbours and this was supported by Malians who preferred local rice.

In addition to the structural changes discussed above, important food culture shifts were also happening at the household level in Bamako. Our surveys with household cooks in Bamako in 2009 revealed that urban families were eating more rice in the run-up to the global food crisis of 2007–08 for at least three reasons (some of which I alluded to in Chapter 5 on Burkina Faso). First, rice had become associated with modernity. This association and attitude had been pushed by the French in the colonial period who pitched rice as a more "civilized" food, but then the idea continued to be fostered (deliberately or not) as rice was emphasized in agricultural development schemes in the postcolonial period. As a consequence, if you were an urban dweller of a certain standing in

Mali, then the expectation was that you should serve rice at the big, noon-time meal. Second, women saw rice as quicker and easier to prepare as compared to traditional grains such as sorghum and millet. This labour-saving aspect of rice was important as urban women often had more demands on their time, leaving fewer hours for meal preparation. Lastly, women appreciated how rice expanded greatly when you cooked it, leaving family members feeling sated after the meal. While rice may have been more expensive per kilo than sorghum or millet at the market, this cost difference was not so great when you compared the end products.

Rising rice prices were a problem in 2007–08 because so many West African urban dwellers were now consuming rice for the structural and household level reasons discussed above. When social unrest broke out in several cities in the region, with citizens protesting against the policies of their governments that had made them so dependent on food imports, there was relative calm in Mali (Moseley et al. 2010; Moseley 2011). There are at least two interesting facets of this unrest, one discursive and the other material. Discursively, this unrest is often tagged as a food riot. The "food riot" nomenclature is problematic because it obfuscates or discounts the agency of the protesters involved. It implies that these demonstrations are spontaneous acts fuelled by anger and hunger. In other words, they are irrational mobs. In my opinion, food demonstration is a much better descriptor because it highlights the agency of the protestors who are fighting for a more just food system. These are political acts aimed at getting the attention of policy-makers and they have often been highly effective in this regard. No matter how isolated or autocratic they may be, most leaders sit up and listen when their urban constituents protest about high food prices (Patel & McMichael 2014). The interesting thing about Mali is that there were no food demonstrations, even though these occurred in nearly all of the surrounding countries in this period (such as Burkina Faso discussed in Chapter 5). There at least two material reasons for this. The first is that Mali, despite neoliberal economic reforms, continued to produce more of its own rice, 80 per cent versus, for example, 40 per cent in Côte d'Ivoire and 15 per cent in The Gambia (and 60 per cent for the West Africa region on average) (Moseley et al. 2010). Unlike Burkina Faso (Moseley & Ouedraogo 2022) or Côte d'Ivoire (Becker & Yoboué 2009), Malians also preferred the qualities of local rice and were willing to pay a little bit more for it. The second reason is that Malian food markets also had plentiful stocks of sorghum on hand when the price of rice started to rise in 2007–08. Sorghum in this context is what economists sometimes refer to as an inferior good, something consumers downshift into when the price of the preferred good becomes too expensive. Our research showed that lower income urban households in Bamako definitely downshifted in this period, somewhat easily pivoting to sorghum when rice became too expensive. Why was sorghum so plentiful and affordable in this time period? This is a question I will explore in the next section.

The cotton strike, cotton retreat and upsurge in Sorghum production

Mali's cotton farmers went on strike in 2000 to protest the low prices being paid for their crop. The low producer price was the result of factors at the international and national levels. At the international scale, the United States had been subsidizing its cotton farmers for years and dumping its cotton on international markets. The resulting oversupply of cotton had led to artificially low prices. A group of Global South cotton producers (including Mali) would eventually and successfully challenge the USA in 2006 in a dispute panel of the World Trade Organization (WTO) for breaking subsidy rules. However, the USA would then (cynically) sidestep the situation by switching its support for American cotton farmers from one type of subsidy to another that was considered legal (Ledermann & Moseley 2007). At the national scale, the Malian cotton company (CMDT) was also benefitting from a difference between the price it paid its farmers for their cotton and the price at which it sold the cotton on the international market (the difference being used to fund CMDT operations and the Malian treasury). In order to buffer against global cotton price fluctuation, the CMDT did have a reserve fund to protect itself and farmers against global price drops. The problem in 2000 was that this fund had been depleted and the CMDT was teetering on bankruptcy, so they had to drop the producer price for cotton to 150 CFA francs per kilo, down significantly from the previous year. As a result, Malian cotton farmers went on strike during the farming season of 2000. Some 50 per cent of cotton farmers would refuse to grow the crop with a 50 per cent drop in production (see Figure 8.1) (Roy 2010).

The story behind the strike is interesting because it was not led by the cotton farmers union known as Syndicat de Producteurs de Coton et Vivriers (SYCOV) for two reasons. First, some would argue that SYCOV was too close to the CMDT and that they had adopted the cotton company's view of the situation. Second, SYCOV, with a stronghold in the Koutiala region, had attempted a strike the previous year and they were bitter that other regions had not followed their lead, thus opting to sit out the strike in 2000. The problem was that many farmers and their village level associations had become deeply indebted following years of low cotton prices. They were desperate. Thus, with SYCOV unwilling to call a strike in 2000, farmers in the other big cotton regions (namely Sikasso and Bougoni) formed a crisis committee in the January–May period to organize outside of SYCOV. The committee drew up a list of grievances, including demands to raise the producer price of cotton and to reschedule the debt that village associations

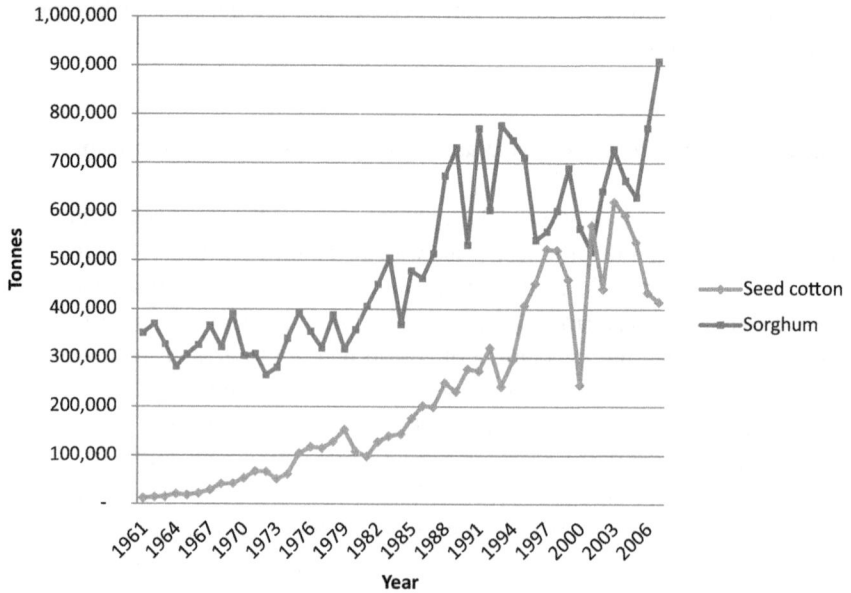

Figure 8.1 Cotton and sorghum production in Mali in the runup to the 2007–08 global food crisis
Source: Graph created by author using World Bank data (2024).

had accumulated.[3] The Malian state and the CMDT refused to negotiate with the cotton farmers and the crisis committee, often referring to them in derogatory terms, such as *broussards* (roughly translated as country bumpkins), who did not understand the situation. The year 2000 was a disastrous one for the cotton company and the government in terms of lost revenues. As a result, they did raise the producer price in 2001 and many farmers returned to producing cotton (Roy 2010). Such a large-scale strike was momentous and, over the longer term, it did foster an atmosphere where farmers were less likely to grow cotton if the conditions were not favourable. In other words, it was a moment of empowerment with longer-term, positive implications for farmer agency, a critical component of the six dimensions of food security discussed in Chapter 3.

---

3. The CMDT had a collectivized debt system. Rather than inputs (seeds, fertilizers and pesticides) being provided on credit to the individual farmers, they were given to the village on credit. As such, more productive and debt free farmers were often covering the debts of those who were not earning enough to cover their input costs. This led to a lot of problems at the village level, including the fracturing of farmer associations into smaller groups of prosperous and not so prosperous farmers (Lacy 2008).

In the years following the strike, cotton production returned to normal, peaking in 2004–05 at 564,971 metric tonnes. But then production began to slide again as the producer price declined. By 2007–08, the cotton crop was down to 240,237 metric tonnes, a decline of over 50 per cent. The producer price in 2007–08 was also 14 per cent below the 5-year average. It seems that the strike in 2000 had opened up a new world of possibilities for Malian farmers as they were now less willing to grow cotton if the conditions were unfavourable and, perhaps most importantly, they were willing to grow more food crops when they were not growing cotton, namely sorghum and maize (Laris *et al.* 2015). As such, farmers pulling back from cotton production allowed for new experimentation in food sovereignty and agroecology, particularly with the traditional grain sorghum. This return to sorghum was simultaneously planned and unplanned. It was planned because some farmers in Mali were thinking about food sovereignty as evidenced by the Nyéléni Declaration, but also unplanned because the cotton strike at the start of the decade had fostered a willingness to pull back from cotton. As evidenced by the trends in Figure 8.1, sorghum production surged in Mali as cotton production declined.

The upsurge in sorghum production allowed Mali's urban consumers and food deficit rural households to weather the 2007–08 global food crisis as they could easily find sorghum at an affordable price (insuring access, one of the key dimensions of food security) when global prices for rice surged (Moseley *et al.* 2010). It is also important to note that Mali's urban consumers, as opposed to urban consumers in other West African cities, were not averse to eating sorghum. While Mali was rapidly urbanizing, this was a relatively recent process, and many urbanites grew up in rural areas and were accustomed to eating sorghum. In other words, sorghum was not considered a retrograde "peasant" grain in Bamako as it might be in Dakar or Abidjan (Delgado & Miller 1985). The other interesting thing to note about declining cotton production and rising grain production in this period was that it contradicted a long-standing adage asserted by Mali's agricultural extension services (touched on in Chapter 1), namely that cash crop and food crop production were complementary. Our own research at the household level had shown that this was only true for the wealthiest farming households, whereas poor and middle income farmers faced clear trade-offs. For these farmers, if they grew more cotton, they grew fewer food crops, a scenario which frequently led to seasonal food insecurity before the next harvest (Moseley 2001). What was interesting about the period in the run-up to the global food crisis (see Figure 8.1) was that this trade-off was reflected in national production statistics, less cotton meant more sorghum and vice versa.

The agroecological benefits of Sorghum production

Sorghum has been farmed in Mali for millennia, was originally domesticated in north central Africa and is well adapted to local environmental conditions (de Wet & Huckabay 1967). Both sorghum and pearl millet are cereal crops in the *Poaceae* grass family and are traditional grain crops grown in the Sudano-Sahel tropical grasslands of West Africa. Pearl millet is the hardier of the two and is grown in areas with 200–800 mm of rainfall per annum (but is more common in the lower rainfall area), whereas sorghum requires more rainfall at 500–1,000 mm per annum and is also more productive in terms of output per hectare. While sorghum is now being replaced by shorter season maize varieties that are more responsive to inorganic fertilizer inputs, and also less nutritious (discussed in Chapter 4), sorghum remains a major grain crop for poorer farmers. The sorghum flour is typically boiled into a stiff porridge, formed into small paddies,[4] and served with a nutrient dense sauce (a baobab leaf sauce rich in iron is common in southern Mali).

One of the great benefits of sorghum is that it works extremely well with agroecological strategies, which makes sense given that it was a mainstay in the traditional food system. Sorghum is often mixed with other crops in the same field, an approach known as intercropping or polycropping. As discussed in Chapter 3, intercropping cuts down on insect predation, helps maintain soil fertility, spreads out labour demands and maintains greater ground cover (which suppresses weeds and reduces water-borne erosion). A very common combination is sorghum and cowpeas (see Figure 8.2). Cowpeas are a legume that fix nitrogen used by the sorghum plant. The plants also do not compete with each other given their very different structures.

Sorghum is also less taxing on the soil than other food crops (such as maize) and grows well without inorganic fertilizer inputs. As discussed previously, sorghum is also drought tolerant and does well under highly variable rainfall conditions. In order to enhance this resilience, sometimes farmers will intercrop various sorghum and millet varieties in order to ensure that there will always be some production.

Political insecurity and food sovereignty

Mali has been wracked by political insecurity since 2012 when Amadou Toumani Toure (aka ATT) was toppled in a coup near the end of his second term. Since that time, Mali has experienced a string of military and civilian leaders (with

---

4. Not unlike Italian polenta that some readers may know.

Figure 8.2 Sorghum and cowpea intercropped in southern Mali
Source: Photo by author.

two coups in the 2020–22 period), an ongoing Islamist insurgency, an attempted break-away state in the northern part of the country, the arrival and departure of the French military, the arrival of Russian mercenaries, human rights atrocities on the part of insurgency groups and the Malian military, and the departure of many aid groups and forms of international assistance (Moseley 2017c; Moseley & Hoffman 2017).

The challenge with such a weak state, insecurity and conflict is that it is extremely risky to develop, sustain and maintain a food system that is premised on exporting cash crops, distributing cash crop earnings and importing food crops. While such systems were already vulnerable to international price fluctuations and supply chain disruptions (Clapp & Moseley 2020), local level conflict and weak governance makes the situation even more untenable. With conflict as a leading cause of food insecurity in the world today, this situation is not isolated to Mali. Furthermore, a growing body of research suggests that agroecology is a sound approach in situations involving weak governance, geopolitical isolation and/or conflict (Moseley 2022c; Tamariz & Baumann 2022). Given the above, Mali's somewhat unplanned experiment with food sovereignty, and the upsurge in sorghum production that transpired when Mali's farmers backed off cotton production, may be worth revisiting. Unfortunately, the current Malian regime of Colonial Assimi Goïta does not appear to be headed in this direction as they have initiated significant subsidies for cotton crop inputs (seeds, fertilizers and pesticides) and cotton production has surged over the past two years (Africanews 2022). While this strategy may be bringing in revenues to the Malian state, it is risky for local food security.

While growing more food locally might appear to make farmers more vulnerable to attacks and looting, the reality is that most attacks are occurring in certain parts of the country and are aimed at high value targets. Most of the farm looting that does occur tends to be focused on livestock theft because it is mobile and nutrient dense (Moseley 2017c). In addition to growing more food locally using agroecological methods (and without expensive, external inputs), cereal banks are another strategy that may be worth implementing more widely under current conditions. The basic premise of a cereal or grain bank is to create a local institution, or buyers' co-operative, that purchases grain after the harvest when it is most affordable and then sells it back to community members with a very small mark-up during the three to four months before the next harvest known as the hungry season (Moseley 1995). Cereal banks help address seasonal food insecurity and cut-out predatory middle-men (grain traders who can earn excessive profits from this situation by buying grain at low prices from desperate farmers after the harvest and selling it back to hungry households at huge markups a few months later). When coupled with increased local and agroecological food production, a grain bank helps a community better address at least five of the six dimensions of food security discussed in Chapter 3 (namely availability, access, stability, sustainability and agency).

Conclusion

This chapter has presented and analysed Mali's somewhat unplanned experiment with food sovereignty in the mid-2000s. As a way to manage risk, Mali's farmers have a longstanding interest in producing at least some of their own food. Malian farmers have also become increasingly well organized, demonstrating an awareness of their own agency within the food system. As a consequence, when Malian farmers pulled back from cotton production in the mid-2000s they also increased the area devoted to sorghum and there was an upsurge in the production of this grain. This surfeit of sorghum, combined with local rice and urban Malians' preference for local grains, meant the Mali rode out the 2007–08 global food crisis much better than surrounding countries. There are important lessons to be learned from this experience, especially given recent insecurity, weak governance and conflict in Mali. A more inward focused food system employing agroecological methods, locally appropriate grains, and cereal banks may make for a more robust, resilient and sustainable food system.

9

BURKINA FASO: PRIVILEGING FOOD SYSTEMS THINKING

Once when I was in our neighbourhood grocery store in a metropolitan area of 2.5 million people in the central United States, I bumped into an anthropology colleague from my university who smiled and wished me well in my foraging. It took me a few minutes to comprehend what he had said and why. What he was implying was that I was interacting with my urban food environment in the same way that a traditional hunter and gatherer might move through a rural landscape foraging for edible fruits, leaves and nuts. While I lived in a built environment, it was nonetheless a food environment, the area from which I sourced my food. I knew the various contours of this environment, including my way around the aforementioned grocery store, the best shop to find the bread our family liked, or the farmers' market where we could source the tastiest sweetcorn in August. Of course, all food environments are not equal. Just a few neighbourhoods to the north of me, in a lower income area, it was harder to find a grocery store, and those shops that did exist often had less healthy food to offer their customers.

Just as I navigated my own food environment in urban North America, Salimata navigated hers in rural, southwestern Burkina Faso.[1] As she walked out to her fields in early July following a rain shower the night before, she noticed that the shea trees (*Butyrospermum parkii*) had dropped some of their fruits over the evening. She spent the next half an hour gathering these fruits under the trees scattered throughout her husband's fields (in a form of parkland agroforestry[2]) as well as from under the trees in the nearby fallow fields that the family had put to rest last

1. This scenario is based on interviews with rural women, and related observations, made in June–July 2017 and June–July 2019.
2. Agroforestry, or the mixing of trees and crops, was discussed in Chapter 3 and is common in many African farming systems. Parkland agroforestry is a typical form of agroforestry and involves leaving desirable tree species scattered throughout the field (for economic and/or ecological reasons) when it is cleared for farming (see a short video I made of agroforestry in Burkina Faso; https://youtu.be/qGjwE9PbviE?si=o8HWq834xzhp0ATQ).

year. While she and her children would eat some of the fleshy shea fruit, the real prize was the nut at the centre of the fruit that she would eventually roast and then crush for its oil. She would use some of this oil, or shea butter, for her own cooking, but much of it she would sell at the local, rural market to traders who drove out from the city of Bobo Dioulasso every five days to buy from women sellers at this periodic market. While gathering the shea fruit, Salimata also noticed that the trees were teeming with caterpillars they referred to locally as *situmu* (*Cirina butyrospermi*). These were a prize that her children adored and they were rich in protein, a treat that was especially welcome at a time of year when food supplies were running short. These she would gather on her way back home several hours later when she was returning from the fields. They would be fried in oil and then consumed as a snack before the evening meal a few hours later. Just as my access to food in my urban food environment in North America was better than that of my lower-income neighbours, so it was for Salimata. As her husband's family was one of the founding families of the village, they had better access to land than those families who had migrated to the village later in time. This was important as tree tenure was based on who held the usufruct rights, or use rights,[3] to the land. Furthermore, as the first wife of her husband in this polygynous society, she had first rights to the products of the trees in her husband's fields.

This chapter explores three inter-related themes as alternatives to mainstream agricultural development thinking. The first concerns the related concepts of food system and food environment. Second, an interrogation of the idea of agricultural revolutions versus the idea of multiple co-existing systems. Third, foraging as an example of a nutrition source in food environments that is often invisible to most agricultural development professionals as well an example of an older approach that has persisted in the face of subsequent agricultural revolutions. I argue that a more holistic food systems approach is critical for improving African food security.

Food systems and food environments

In contrast to the linear value chain model privileged by the New Green Revolution for Africa framework (discussed in Chapter 5 on Burkina Faso),

---

3. Usufruct rights, or use rights, were discussed in Chapter 1. They refer to the tenure rights that households' have to land in a common property system. These rights are recognized within the community, but such property may not normally be bought and sold as is done in a private property regime. Tree tenure is also typical in common property systems. Those who hold the use rights to the land also often hold the tree tenure rights, or the rights to harvest the fruits of the trees.

a focus on food systems and food environments encourages governments and development organizations to consider the broader system and environment in which a farm field and household is located. In other words, as Gengenbach *et al.* (2018) argue, AGRA's value chain model is an oversimplification and it makes more sense to think of local food systems as complex assemblages. These concepts are also highly consistent with agroecology which not only pushes us to consider the ecological interactions occurring on a farm, but also the broader ecological and social landscape in which the farm is located. For food systems in particular, there are also strong links with political ecology which considers the power dynamics and linkages between different actors in the food system.

A food system is the assemblage of actors that provision any given area with food. This includes all of the actors from the point of food production to the point of consumption, including production, processing, distribution, and consumption. This ecosystem of actors is further influenced by inputs such labour and energy, as well as waste created at different steps. Furthermore, this system is shaped by policy and climatic conditions. While some aspects of this conception overlap with the value chain concept discussed in Chapter 5, a food system involves multiple value chains and is often thought of as a web rather than a chain. In some ways, this is the human-inflected version of the food web concept in ecology. As discussed in the introduction, food environments are the areas in which people live and source their food, whether they live in urban or rural areas (HLPE 2020). Some food environments are richer than others and many aspects of the food environment may be illegible to external development actors. This can lead to a "bull in the china shop" problem wherein development projects unwittingly compromise (or destroy aspects of) food environments if they are not careful, especially common pool resources that poor households may depend on. For example, in my Burkina Faso study in 2019 I saw vast new irrigated rice perimeters, with large diesel motor pumps, being constructed along the Mouhoun River in southwestern Burkina Faso. On the one hand, such a major investment looked like a huge advancement for rural development in the area. On the other hand, I wondered what the consequences would be for those who traditionally foraged in the seasonally flooded areas along the river, or who would actually get plots inside the perimeter, or who could afford the obligatory dues that would be levied to pay for the diesel fuel to run the irrigation pumps.

So what does the typical food environment and farmscape look like in rural, southwestern Burkina Faso (see Figure 9.1)? Many farms in this area are strategically and non-contiguously spread across a mosaic of forests, fields and grasslands. As discussed earlier, the village territory is managed as a commons with households controlling the use rights to fields emanating out from the village centre where most family compounds are located. Households may control fields in different microenvironments, such as low-lying areas that flood in the rainy

Figure 9.1 Local food environment with field and agroforestry in the foreground and village in the background
Source: Photo by author.

season (ideal for rice) or upland areas that are better for other types of crops, such as sorghum, maize and cotton. This patchwork of fields is bisected by fallow fields, gallery forests, water courses, uncultivated hilly areas and pastures. This village territory abuts other village territories, villages are connected by paths and roads, and often times larger villages and towns host weekly markets where producers and vendors periodically gather to sell everything from staple grains, garden vegetable and fruits, to clothing, cookware, and candies, to pesticides, fertilizers and seeds. Traders from urban areas also drive out to these markets to buy up rural produce for sale in the city or to destinations further away. Households live in relation to this food environment, procuring food from all types of sources: crops from their fields; vegetables from their gardens; wild tree products they collect; fish they catch in the stream; pasta and seasonings they purchase at the weekly market; eggs from their guinea fowl that range across this landscape; or the sack of rice that their relative in the capital city sends home every two months. It is this entire food environment that nourishes a rural family rather than only a farm field tilled for certain crops and a market where some food stuffs are purchased. A lack of awareness of this larger food environment, and the variety of foods it offers, may also lead to the development of one source of food at the expense of another. For example, pesticide runoff from farm fields may harm fish populations in seasonal streams, or the clearing of new areas for farming may lead to loss of trees that provided valuable fruits, nuts and leaves in nutrient dense sauces made to accompany a cooked staple grain at the noon meal.

Challenging the narrative of agricultural revolutions with that of multiple systems

Traditionally we teach that three major agricultural revolutions have occurred alongside the economic development of human societies (Fouberg & Moseley 2018). As human population densities increased, and foraging resources became more scarce, we see the first agricultural revolution, or the idea that people gradually moved from hunting and gathering to crop farming (Diamond 1987). This first happened around 10,000 years ago in Mesopotamia, or the area between the Tigres and Euphrates Rivers in modern day Iraq, but then gradually spread to, or independently happened in, other parts of the world.

Then as human population densities increased further, there was a shift from extensive agriculture and long fallows, to intensive organic agriculture involving manure and compost and shorter fallows, or the second agricultural revolution in the eighteenth century. Higher yields would have allowed for some level of surplus production and urbanization. This likely first happened in the most densely populated parts of the world, such as some parts of China and Europe, and then spread. While not conventionally framed as such, I would also argue that the first and second agricultural revolutions were also the cradles of agroecology as many of the techniques associated with agroecology would likely have emerged during these transitions. Agroforestry was likely part of the first agricultural revolution as people no doubt left some trees that were valuable to them in the fields they created. Manuring and composting, which are part and parcel of agroecology, were key innovations of the second agricultural revolution. To this I would add polycropping, which likely increased in importance during the second agricultural revolution.

As nation states became more urbanized, they also fought over organic fertilizer resources, such as the Guano Wars, (Moseley *et al.* 2013) where Europeans and Americans fought over small islands with centuries of accumulated bird droppings, a resource that was mined and shipped back to the home country. The discovery of how to synthesize nitrogen, known as the Haber-Bosch process in the nineteenth century,[4] led to the development of inorganic fertilizers, freeing farmers from somewhat finite organic fertilizer resources. This, combined with a move beyond traditional plant breeding and the development of hybrid seeds[5] that were more responsive to inorganic fertilizers, led to the third agricultural revolution in the late nineteenth/early twentieth century in the United States. This third agricultural revolution, involving inorganic fertilizers, hybrid seeds and mechanization, was also much more capital intensive. To this suite of more capital-intensive inputs, would be added increasing use of pesticides after the Second World War. The third agricultural revolution was a boon to agribusiness and the science of agronomy, which became increasingly intertwined, as this approach evolved over the twentieth century. This approach spread to Europe and Japan, and to white settler colonies such as Australia, New Zealand and South Africa as a function of little d development. The big push to introduce this to the Global South as a form of big D development[6] came in successive waves of the green revolution, the first wave during the 1950s–70s, and the new Green Revolution for Africa in the 2006–20 period (discussed in Chapter 2).

The problem with the idea of agricultural revolutions is that it is presented as a paradigm shift wherein one form of agriculture completely supplants another. Furthermore, it is often conceptualized as a linear model where each new shift or revolution is framed as a form of progress. Based on what is happening on the ground in southwestern Burkina Faso, and in other parts of Africa, this is not an accurate description. Rather, farmers are diversifying their approaches, experimenting with some new ideas but also retaining many aspects of older systems (Moseley 2018; Morgan & Moseley 2020). In other words, African farmers are often expanding their repertoires, not moving from one system to the next. While colonial agencies, agronomists and development agencies have often framed this syncretic approach as problematic and as peasant backwardness

---

4. "The Haber-Bosch process is a process that fixes nitrogen with hydrogen to produce ammonia – a critical part in the manufacture of plant fertilizers. The process was developed in the early 1900s by Fritz Haber and was later modified to become an industrial process to make fertilizers by Carl Bosch. The Haber-Bosch process is considered by many scientists and scholars as one of the most important technological advances of the 20th century" (Briney 2019).

5. Traditional plant breeding, based on the work of Mendel, involved cross pollinating plants with desirable characteristics (and physically placing the pollen of one plant on another). Hybridization entailed inbreeding two inbred plants with the same desirable characteristic (e.g. bigger grain seeds or shorter stalks) for the hyper expression of that characteristic. The downside of hybrids for farmers is that you have to buy new seeds every year because the second generation seeds they produce are far less productive. However, this shortcoming for farmers is a boon to capitalists as it guarantees an annual market for the seeds they develop (Tripp 2001).

6. The concepts of big D and little d development are introduced in Chapter 1 and are based on the work of Vicky Lawson (2007).

or resistance to change, it is clearly a way to manage risk and build resilience. Of course, money, power and force are behind the third agricultural revolution in ways that never existed for the first and second agricultural revolutions, but rural people are adept at side stepping risky propositions as evidenced by the continued relevance of foraging and full use of rural food environments in southwestern Burkina Faso.

Foraging

I first came to understand the importance of foraging (that is, the collection of fruits, leaves and insects from wild and semi-wild plants and trees) when I ran a large food security project in the mid-1990s in central Mali for an international NGO called Save the Children UK. There foraging repeatedly came up as an important coping strategy in lean years or during the hungry season (Moseley 1995; Davies 2016). But in southwestern Burkina Faso, it was really the insistence of my former student Julia Morgan that persuaded me that this was an important topic worthy of investigation (Morgan 2018; Morgan & Moseley 2020). During surveys we conducted in 2017 and 2019 (both after the harvest and during the hungry season), we asked women farmers what they and their families had consumed over the past 24 hours. These results showed that my student Julia was on to something as different foraged foods kept showing up in our surveys, particularly during the rainy season. Another student, Jane Servin, and I were able to further elaborate on the initial work Julia had done. Table 9.1 shares some of the foraged foods mentioned by women in our surveys whereas Figure 9.2 shares information on how commonly these foods were mentioned by interviewees. Clearly, large proportions of the women farmers we were interviewing were also engaging in foraging and this held up across difference wealth groups (it was not just a survival mechanism for the poor).

Some discussion of a few specific examples of foraged foods is useful for understanding their position in local food environments and their place in diets. Here I talk about two: African baobab and African locust bean. African baobab (*Adansonia digitata*) is a naturally occurring species across the tropical African grasslands, both in the northern and southern hemispheres. Historically it was propagated by elephants who ate baobab fruits and scattered the seeds with their faeces (Napier-Bax & Sheldrick 1963). A similar role has been played by chimpanzees in West Africa (Duvall 2007). The hard seed or pit at the centre of the baobab fruit will only germinate after its protective covering has been removed by an elephant or chimpanzees' digestive tract. In areas where elephant and/or chimpanzee numbers have declined, humans have taken to preparing the seeds and planting them. Baobab trees, which are known for their distinctive structure (some say it

Table 9.1 Scientific and common names for wild and semi-wild foods recorded in surveys

Scientific name	English name	Dioula name	Mooré name	Product	Dietary diversity categories
Adansonia digitata	African baobab	Zirasun	Toega	Leaves	Dark green leafy vegetables
Bombax costatum	Red-flowered silk cotton tree	Bambou	Vuaka	Fruit	Other Fruits
Ceiba Pentandra	Kapok	Bana yiri	Gounga	Leaves	Dark green leafy vegetables
Ceratotheca sesamoides	False Sesame	Banougou	Boundou	Leaves	Dark green leafy vegetables
Cirina butyrospermi	Shea caterpillar	Shitumu	UK	Insect	Other foods and beverages
Corchorus olitorius	West African sorrel	Fonongoh	Bulvaka	Leaves	Dark green leafy vegetables
Mangifera indica	Mango	Mangue	Mangue	Fruit	Vitamin A Rich fruits
Parkia biglobosa	African locust bean	Nèrè	Roànga	Fruit, Seeds	Fruit: Other fruits Seeds: Legumes, nuts, & seeds Leaves: Dark green leafy vegetables
Vitex doniana	Black plum	Koto	Anda	Fruit, Leaves	Fruit: Other fruits Leaves: Dark green leafy vegetables
Vitellaria paradoxa	Shea	Shi	Taanga	Fruit, Seeds	Fruit: Other fruits Seeds: Oils & fats

Note: Spelling for Dioula and Mooré are phonetic. UK = Unknown.
Source: Morgan (2018).

looks like a tree planted upside down) may be found scattered around villages, in farm fields and in shrublands (see Figure 9.3). Sometimes baobab groves in the middle of shrubland areas suggest that it was a previous site of human habitation as the trees are known to live for hundreds of years (beyond the lifespan of traditional adobe construction) (Duvall 2007). It is the baobab tree leaves which are harvested as food. These leaves are dark green and incredibly rich in iron, which is especially important for women of reproductive age. Younger women will often scale the trees (and they are somewhat easier to climb given their structure) and pick the leaves that are needed for a meal. Given tree tenure arrangements, women know the trees from which they may harvest leaves or not. The leaves may also be dried and used at a later date. Baobab leaves are a very common ingredient in local sauces and may accompany the main rice, sorghum or maize-based starch.

African locust bean (*Parkia biglobosa*) or *néré* is another naturally occurring tree species which self-propagates. These too may be found scattered throughout

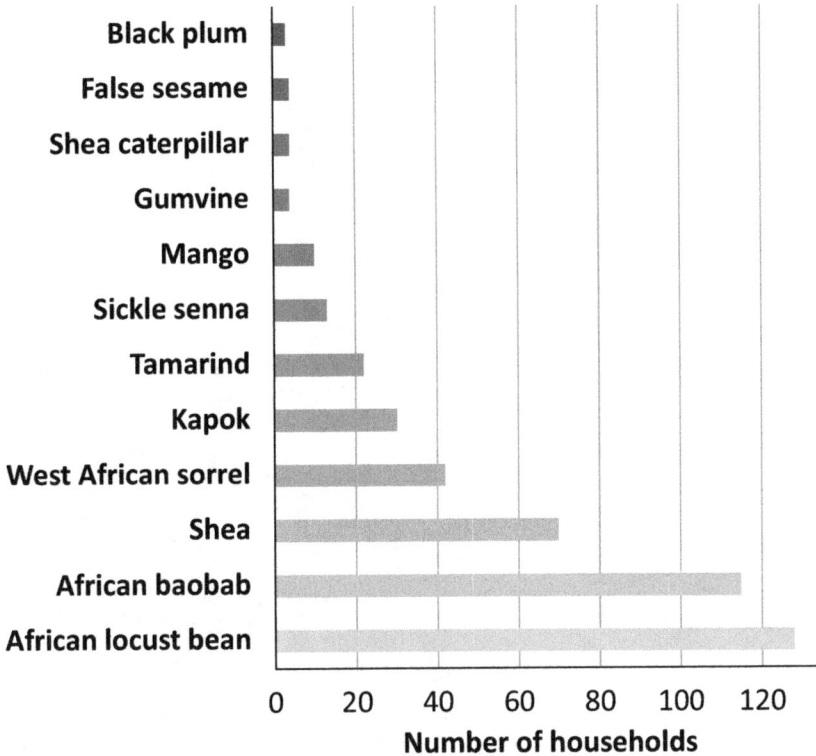

Figure 9.2 Number of participants who ate each wild or semi-wild food at least once in the four rounds of data collection. The maximum is 131.
Source: Servin and Moseley (2023).

shrublands and farm fields, but they are more widely dispersed and thus easier to harvest from in common areas where tree tenure is not as restrictive. Women collect the pods by knocking them down with large poles (as the trees are more difficult to scale) or collecting them off the ground. Women will store these pods on the roofs of houses in order to keep them away from animals. When they are ready to prepare them, they extract the seeds from the pods and ferment them, turning them into a nutrient dense sauce ingredient known as *soumbala* (which is a good source of protein and other nutrients (Termote *et al.* 2022). *Soumbala* has a distinct, fermented smell that is sometimes off putting to the uninitiated, but a real favourite of locals.

In Burkina Faso, my students, collaborators and I came to understand that foraged foods were incredibly important for dietary diversity (Morgan & Moseley 2020; Servin & Moseley 2023; Tkaczyk & Moseley 2023). More specifically, we found that increased foraged food consumption was associated with higher measures of food security (as measured by the household food insecurity access

Figure 9.3 African baobab (*Adansonia digitata*) in southwestern Burkina Faso
Source: Photo by author.

scale) and dietary diversity after adjusting for wealth (Morgan & Moseley 2020).
That said, wealthier households tended to have better dietary diversity scores
than poorer households, and foraged foods tended to play a bigger role in dietary
diversity for the poor as compared to wealthier households (Servin & Moseley

2023). In other words, foraged foods were important for the dietary diversity of all groups in this rural area, but more important for the poor. We also found that local food environments mattered and helped explain differences in foraged food consumption across the survey area (Servin & Moseley 2023). Lastly, higher indices of food consumption agency (which was based on questions about people's ability to act on their food preferences and level of worry about future meals), tended to be correlated with higher levels of dietary diversity and foraged food consumption. That is, people preferred foraged foods and would act on that preference if they were able. Interestingly, higher household wealth did not correlate with higher measures of food consumption agency (Tkaczyk & Moseley 2023).

Putting all of above together, we found that wealthier households in rural southwestern Burkina Faso did eat better than poorer households. Furthermore, foraged foods were important contributors to dietary diversity across the rural wealth spectrum but more important for the poor than the wealthy. Lastly, people liked foraged foods and would eat them if they could. This means that foraged foods are simultaneously a seasonal treat and a nutritional safely net (which goes against some characterizations of foraged food as less desirable famine food) (Huss-Ashmore & Johnston 1997). Lastly, local food environments matter, which explains some of the variation in foraging across villages in the survey data.

Conclusion

In conclusion, the experience in southwestern Burkina Faso suggests that we need to think beyond the farm field and about the larger food environment and food system. While we sometimes think of foraging as a relic of the past that has been supplanted by subsequent agricultural revolutions, or only something rural dwellers revert to out of desperation, the reality is that it never went away and has continued to exist alongside other forms of food production. Furthermore, this almost invisible form of food procurement is more ubiquitous than we think and fulfils a really important role in terms of dietary diversity and addressing micronutrient deficiencies. Lastly, in Burkina Faso, foraged foods are especially important during the so-called hungry season, a time before the next harvest when many rural families are running low on grain and cash. This is also a time when people are working hard in the fields during the rainy season, serendipitously a period when many wild plants are leafing out or fruiting.

To be clear, I am not arguing for an all-out return to foraging based on this analysis. Rather, I think it is important to recognize that this form of food procurement persists and that it is not some anachronistic relic of an older system. Furthermore, it is an ecologically efficient and non-destructive way of harvesting food from the landscape. While it may not fulfil a significant proportion of total caloric needs, it is important for providing certain micronutrients and an important way to address micronutrient deficiencies, which is a form of malnutrition that often gets less attention.

The policy implications of this case of an under-the-radar, or hidden food safety net in southwestern Burkina Faso are at least two. The first lesson is to be more fully aware of the opportunity costs of blindly pushing new forms of agricultural development if they force out existing forms of food procurement, such as foraging and more agroecological cropping systems. The second implication is that the idea of multiple, co-existing food procurement systems offers a concrete alternative to the zero-sum conception of parks and preservation areas versus agricultural sacrifice zones (Terborgh & Van Schaik 2002). The latter is sometimes referred to as the land sparring argument wherein one must intensify food production using improved seeds, pesticides and fertilizers so that one may set aside land for preservation (Perfecto & Vandermeer 2010). The problem with agricultural sacrifice zones is that they are ecologically thin, or less resilient, and they do not feed the food insecure. Parks in this context are also often not supported by the local population because they lost land, or access to resources, to create these preservation areas. If western conceptions of preservation as a zone where no resources are harvested (Cronon 1996) were relaxed, one could easily imagine foraging being allowed in core preservation areas. This would generate good will, provide vital nutrition resources and not negatively impact the basic functioning of the ecosystem. Furthermore, moving out from core preservation areas, one could imagine biodiverse, agroecological farming areas with parkland agroforestry systems. What I am describing has sometimes been labelled as an agricultural matrix (e.g. Vandermeer & Perfecto 2007), or aspects of convivial conservation (Büscher & Fletcher 2019), and the importance of foraging in some African food systems suggests that it would be a better model than fortress conservation for the promotion of biodiversity.

10

GENDER MATTERS: WOMEN FARMERS, WATER AND CLIMATE CHANGE IN BOTSWANA

I spent a semester teaching at the University of Botswana, a large public university in the capital city, Gaborone. My family and I lived in a nearby residential area and we became friendly with many of the neighbours in our apartment complex, made easier because our children frequently played in the communal courtyard. One weekend, the neighbour across the way from us invited us to go see his family home and mother who lived about a two-hour drive to the north of the capital. Botswana is one of the least densely settled countries in the world, so driving north into the drylands surrounding the Kalahari Desert meant passing through a lot of semi-arid shrubland, with only the occasional sight of people outside of a town as we drove there.

We spent a lovely day at their home, eating, chatting, playing with the children and walking around the village. We then piled into the car and drove out to the mother's farm fields, which were some distance from the village centre as is common in Botswana's rural spatial arrangements (see Chapter 6). We walked all over her fields, saw what she was planting, and chatted with her at length about her challenges as a woman farmer in Botswana. What really struck me at the time was that she was pretty much doing it all on her own. All of the men in the family had moved to the city, including her son (our neighbour) and his brother. While the mother clearly benefitted from the remittance income her sons provided her, she also did not have enough labour to run her farm. What also struck me was her discussion of increasingly irregular rains and the regular loss of crops. Her farming livelihood was extremely marginal. Later that same afternoon, we visited her neighbour's cattle post. Even through my American gaze, which had grown up seeing dairy cows in Wisconsin, this man clearly had robust, healthy-looking beef cattle. My children watched with glee as the cattle slurped up water from a trough provisioned by a borehole well. I, however, kept thinking of my friend's mother's crops withering in the field for lack of rain versus these well-watered cattle.

A few years later, my student (Rachel Fehr) and I would return to Botswana to study a backyard gardening initiative and gardening more broadly among women, something I had discovered while conducting research on household reactions to high food prices during my semester teaching at the University of Botswana. One day we were interviewing women gardeners along Ngotwane River north of Gaborone. Here along the river were dozens of women with small gardens and motor pumps, producing plenty of vegetables for their own families, as well as extra to sell at the market. Compared to my neighbour's mother, they had a huge advantage because they had regular access to water. They also intensively farmed relatively small plots, which – while highly productive – cut down on the need for labour.

Access to water, the gendered dimensions of farming, and male biased agricultural policies are some of the key issues facing Botswana's agrifood systems today. This chapter explores a backyard gardening initiative in Botswana that was one of the country's first programmes to overtly support women farmers. While widely condemned as a failure, our research showed that those women horticulturalists who had adequate access to water actually did quite well. I begin by discussing the gendered dimensions of farming in Botswana and feminist political ecology. I then examine the aforementioned backyard gardening initiative. I lastly examine the complicated issue of water and, more importantly, the sustainable use of water resources for agriculture in a semi-arid country.

Women, agriculture and gardening

As noted in previous chapters, Botswana is rapidly urbanizing and is a majority urban country. About 30 per cent of the population lives in rural areas and 52 per cent of this rural population is female (FAO 2018). Women compose a large share of the agricultural workforce and do a significant amount of the crop farming but still often show up as less involved than men. Agricultural statistics broken down by gender in Botswana are somewhat limited (FAO 2018); nonetheless, according to the annual agricultural survey, 36.85 per cent of farms in Botswana are owned by women (Statistics Botswana 2020). Furthermore, as Botswana has urbanized, the share of employment in agriculture has risen for women (from 32 to 39 per cent) and declined for men (from 68 to 61 per cent) (Mackett 2021). Since much of female agricultural work is unpaid, a lot of it goes unrecorded in official statistics. Furthermore, oftentimes a male head of household is listed as the farm owner, even though women do the majority of crop farming work. Both factors reveal women's important yet marginalized role in Botswana's agriculture.

In the first part of this book, I introduced political ecology as a conceptual framework and briefly mentioned a subfield known feminist political ecology. While feminist political ecologists acknowledge that gender is socially constructed, they also know that it has real material consequences in terms of gendered roles and responsibilities (Rocheleau *et al.* 2013). This is especially true in African agriculture where men and women often perform distinct tasks. In Botswana, men were traditionally responsible for the clearing of land and ploughing with cattle, guarding fields, hoeing and doing equipment checks. Women were responsible for the entire crop production cycle, including the planting or broadcasting of seeds, weeding, harvesting, processing and post-harvest handling. For animal husbandry, women generally tended small stock (especially goats) and chickens, and men cared for large livestock, notably cattle. It was also women's responsibility to collect firewood, forage foods (aka veldt products) and fetch water (FAO 2018).

Because of gender disparities, women are often given less productive land or may not have the same access to water for their livestock (with production implications). Urbanization has also had big impacts on the rural farming sector, especially as men leave agriculture. This has contributed to a much larger trend in the Global South known as the feminization of agriculture that has accompanied male migration and urbanization (Slavchevska *et al.* 2019). In Botswana, many tasks traditionally performed by men, such as the clearing of land and ploughing, are no longer performed by them. To wit, the mother of my neighbour (who I presented in the introduction) had trouble finding someone to plough her fields given the absence of men in her rural household. She eventually paid a male farmer with a tractor to do this work for her and the Botswana Department of Agriculture does have a grants programme to support this. While gendered farming roles have often put women farmers at a disadvantage, sometimes it has worked the other way. Alice Hovorka (2006)'s research, for example, demonstrated how women's traditional role raising chickens was able to expand and become more profitable as Botswana urbanized. Broadly, she argues that women may be less marginalized in urban agriculture (where they participate in equal numbers to men) than in more traditional rural agricultural spheres.

Unlike some other African countries, gardening is a relatively new phenomenon in Botswana. As discussed in Chapter 6, the most commonly raised food crops in Botswana are maize, sorghum, millet and pulses (mainly cowpeas, Bambara groundnuts and mung beans) (FAO 2018). A traditional Botswana meal featured sorghum or corn meal porridge. A thicker version of this porridge, known as *bogobe* in Setswana, was the typical midday meal, and was ideally accompanied by a stew of meat and/or vegetables or beans (FAO 2018). The traditional greens in these stews would have been foraged wild greens rather than cultivated ones. The absence of gardening is likely explained by two factors. First, a lack of rainfall that made it hard to grow many vegetables and the privileging of well or bore hole water for animals. And second, the ubiquitous presence of domestic animals on the landscape, and especially around water points, would have made gardens vulnerable to animal incursions (and many male animal owners would likely have refused to compensate women for their vegetable crop loses). But urbanization has created an opening for gardening, and this trend has been defusing out from Gaborone to smaller towns and rural areas. Now the traditional stew accompanying the starch in the midday meal frequently includes cooked cabbage or spinach. While the traditional diet that is benefitting from gardening is quite healthy, the other parallel trend accompanying urbanization is the nutrition transition (discussed in Chapter 6) wherein people are consuming more meat, fat, sugar and processed foods (with associated increases in obesity and non-infectious diseases) (Nnyepi *et al.* 2015).

The backyard gardening initiative

In 2008, the Government of Botswana launched a backyard gardening initiative as part of then president Ian Khama's Poverty Alleviation Programme. This initiative targeted poor households, providing them with water reservoirs, shade netting, seeds, tools and extension services in order to become commercially oriented gardeners (see Figure 10.1). While this was a rare instance of the government supporting a women's agriculture activity, by 2012 the backyard gardening initiative was declared a failure, widely dismissed by the public as a waste of resources, and eventually shut down. For example, at the time the *Sunday Standard* newspaper (2013) reported "backyard gardens going through a rough patch". A more recent (2023) government audit of the backyard gardening initiative found that 37 per cent of the gardens had failed, mostly due to high water bills. In this report, the Botswana government auditor general wrote that "These projects resulted in wasteful expenditure exceeding P2 million" (Tlhankane & Mguni 2023).

I first encountered the backyard gardening initiative when I was doing research on household food insecurity in Botswana in 2012. At that time, I kept running into these gardens when I was doing household surveys in the peri urban areas surrounding Gaborone. Intrigued by this initiative, my student Rachel Fehr and I returned to Botswana in 2015 to study women gardeners more formally. Unlike the government of Botswana, we were less concerned about evaluating these gardens in terms of their commercial success and more interested in their

Figure 10.1 Botswana backyard gardener with shade netting and drip irrigation
Source: Photo by author.

potential food security benefits (although admittedly these are two factors are sometimes linked). More specifically, we sought to better understand the relationship between food security, the commercial orientation of gardeners, and access to stable and affordable supplies of water for gardening. We interviewed 100 women that were split between 25 who gardened for commercial reasons, 25 who gardened partially for commercial and partially for home consumption, 25 who mostly gardened to feed their families, and 25 who did not garden (a control group).[1] These women gardeners also had varying access to water, including private bore holes, access to river water and metered municipal tap water. The women were located in seven different large villages within 60 km of Gaborone (the most densely populated part of Botswana). We measured food security using the household food insecurity access scale (HFIAS) and the Household

---

1. We initially identified some women who had participated in the Ministry of Agriculture (MOA)'s Backyard Gardening Programme and then found others through a snow balling technique. Our sample was skewed towards women who were easier to contact, gardens that were more accessible and gardeners that were known to the MOA.

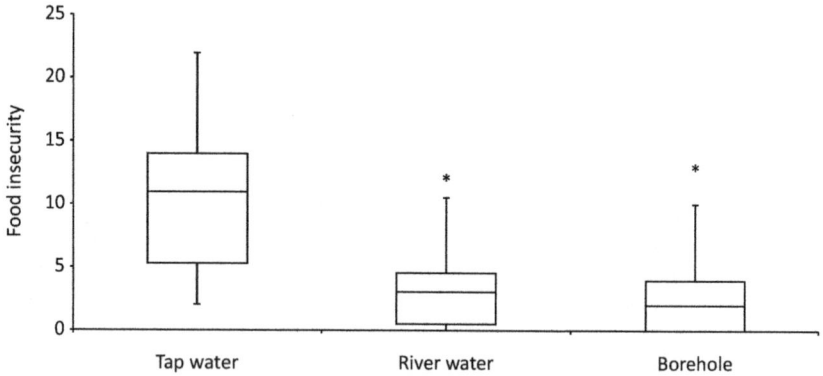

Figure 10.2 Food insecurity among women gardeners by water source
Note: Higher scores indicate more food insecurity.
Source: Fehr and Moseley (2019).

Dietary Diversity scale (HDDS) discussed in Chapters 5 and 6.[2] We used an expanded HFIAS scale (with a range of 0–27, rather than the typical 0–4) to allow for better statistical analysis.[3]

We found that those households with women gardeners that had access to borehole water or river water were both fairly food secure, whereas those that relied on tap water were not (see Figure 10.2) (Fehr & Moseley 2019). More specifically, borehole gardeners had a mean HFIAS score of 2 and river water gardeners 3 (both high food security), whereas tap water gardeners had a mean

---

2. The household food insecurity access scale (HFIAS) is based on the answers to nine qualitative questions about the previous four weeks. These range from more moderate questions (how many days did you worry about not having enough food et eat) to more severe (how many times in the past 4 weeks did you go 24 hours without eating). The score ranges from 0 to 4, with higher scores indicating greater levels of food insecurity (Coates *et al.* 2007). The household dietary diversity score (HDDS) is based on a recall of all the foods household members have consumed over the past 24 hours. These foods and ingredients are then categorized according to major food groups and a score for the total number of food groups is given (with higher scores being better) (Swindale and Bilinsky 2006). While this is a measure of dietary diversity, it is also considered to be a proxy for food security.

3. The nine questions in the HFIAS questionnaire, each scored 0–3, can lead to a total score between 0–27 (from perfect food security to acute hunger). However, this does not allow for the weighting of questions by severity. Thus, for example, a question on worrying about lack of food is weighted the same as going 24 hours without eating. The FAO does have a system for weighting the questions, but it results in an index of 0–4 that is too coarse for statistical analysis (and ideally the dependent variable in statistical analysis will vary substantially). Rachel Fehr (2016) developed an alternative 0–100 scale based on the questions (from perfect food security to acute hunger), but it added little nuance to the unweighted 0–27 HFIAS scale. As such we use the 0–27 scale here.

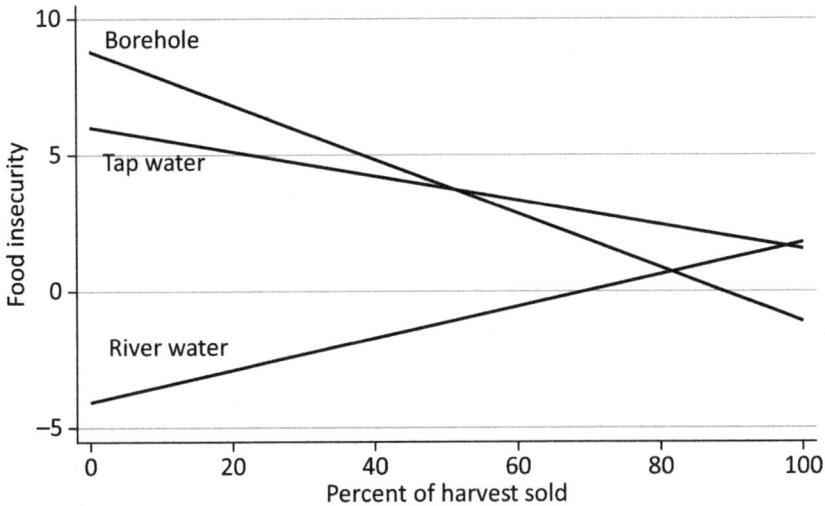

Figure 10.3 Relationship between food insecurity and commercialization by gardener water source
Note: Higher scores indicate more food insecurity.
Source: Fehr and Moseley (2019).

score of 11 (with higher scores being worse). This was also reflected in the house-hold dietary diversity scores, wherein borehole users ate food from, on average, 3.68 different groups, river water users 3.08 different groups, and tap water users 2.59 different groups.[4]

The type of water source had a significant impact on the dynamic between the level of commercialization and food security in our study. These relationships are visually depicted in Figure 10.3. For tap water users, for each 10 per cent increase in portion of harvest sold, there was a 1.11 per cent decrease in food insecurity (however, this relationship was not statistically significant). For borehole water users, there was a stronger relationship. For each 10 per cent increase in portion of harvest sold, there was a statistically significant 24.6 per cent decrease in food insecurity. Lastly, and perhaps most interestingly, for river water users, for each 10 per cent increase in portion of harvest sold, there was a 9.12 per cent increase in food insecurity (and this finding was highly significant).

While these findings may appear odd at first glance, there is a reasonable explanation. First, the tap water users faced the challenge of having to pay for

4. The reader may notice that these HDDS scores are much lower than what I found working in southwestern Burkina Faso, where mean scores ranged from 5.7 to 6.1 (reported in Chapter 5). It could just be that southwestern Burkina Faso has a richer food environment or that more rapid economic growth in Botswana has led to a decline in dietary diversity.

municipal water. As such, they more or less had to sell a good portion of their vegetables in order to cover their costs. The problem was that it was somewhat risky to participate in commercial vegetable markets and the tap water gardeners, who tended to be poorer, did not always navigate these markets very well. As such, those who sold more did better than those who did not, but their overall food security was the lowest of the three groups and was not significantly better than those who did not garden at all. The borehole water users had to make a big investment up front to sink a well for gardening, but then the water they used was relatively inexpensive to access. This group was also a little wealthier and more business savvy, making it less risky for them to engage with commercial vegetable markets. Their situations improved dramatically the more they engaged with the market, allowing them to pay off their borehole loans and bank the remaining profits. The last group, the river water users, were benefitting from their proximity to the river. They did not need to make a major investment in order to access this water, other than the recurring costs of motor pump fuel. For this group, the advantages of selling more and more vegetables were limited (as they did not have major expenses to cover), yet the risks of engaging in commercial vegetable markets were there. As such, the food security benefits for this group were higher when they held back enough food for their families.

In sum, despite the misgivings of the government and the press in Botswana, we would argue that gardening is an important new avenue for women in Botswana that can lead to food security improvements and deserves further public support. What many commentators failed to grasp was that the success of gardeners was highly contingent on access to water. We found that those women with better access to water were highly successful in food security terms (Fehr & Moseley 2019). A greater awareness of the gendered dimensions of conventional agricultural development, which historically focused on the cattle industry in Botswana, creates a space to better address the food security of marginalized groups, including women in Botswana. An additional advantage of gardening is that it supports healthy eating, especially when the vegetables grown can fit into the traditional diet.

Water, climate change and sustainable adaptation

Access to water is a significant challenge in a semi-arid country like Botswana. Worse yet, most downscaled climate change models predict a 50–100 mm decline in rainfall in southern and southeastern Botswana where most people live and a 1.5–2.5°C increase in temperature between now and the year 2050 (Batisani & Yarnal 2010; Zhou *et al.* 2013). Rainfall declines will make more marginal agriculture even more tenuous and temperature rises will increase rates of evapotranspiration. This is a longstanding challenge for semi-arid Botswana that will only become more taxing in the years to come.

The problem is that many conventional approaches to securing water in dry regions are both environmentally and socially problematic. As discussed in Chapter 6, the Government of Botswana has supported the drilling of boreholes throughout the country in order to aid the cattle industry. The borehole strategy is unsustainable for two reasons. First, the proliferation of boreholes has led to a drop in the water table in many places as water is taken out more quickly than it can be replenished. (Manzungu *et al.* 2009). Second, even with government support, boreholes are expensive to drill and maintain. As such, this is a strategy that tends to benefit the wealthy and further marginalize the poor.

Even more complicated are urban water issues. The placement of Botswana's new capital city, Gaborone, after independence was largely driven by concerns about water access. Of the potential locations in southeastern Botswana where the majority of the population resides, Gaborone was chosen because a dam could be built on the Ngotwane River (which is a tributary of the Limpopo, one of southern Africa's most significant rivers) (Keiner *et al.* 2013). The resulting Gaborone dam has been the main source of water for Gaborone since it was constructed in the 1970s. The problem is that as the population of the city of Gaborone grew, and as the region became drier, the reservoir has nearly gone empty on multiple occasions (Arku 2009). In fact, during the winter of 2015 when I was doing research in Botswana, my friends and I climbed to the top of nearby Kgale Hill to look out over the reservoir behind the dam and there was almost no water to be seen (although the dam had filled back up again when I returned in 2019). Of course, several meagre rainy seasons and a growing city had exacerbated the problem, but impounded water is also highly susceptible to evaporation.

Botswana's solution to Gaborone's water problem has been the north–south water carrier, a huge pipeline that brings water from the north to the south of the country (Moseley 2015c, 2016). This was originally built in 2000, and extended some 360 km to the northeast to tap reservoirs built along other tributaries of the Limpopo River (near Francistown along the border with Zimbabwe) (see Figure 10.4). A further extension of the pipeline was completed in 2023, tapping into a dam on the Shashe River, which also flows into the Limpopo (Morwaeng 2023). Now a third phase of the pipeline is envisioned that would add another 500 km and go all the way up to the town of Kasane (near a confluence of borders of Zambia, Zimbabwe, Botswana and Namibia) and bring down water from the Zambezi River (another major, yet completely different African river system (GoB 2010)). As a relatively wealthy, middle income country, Botswana can afford to take an engineering approach to address its water scarcity challenges. However, as we know from cities like Los Angeles, Phoenix and Riyad, this can

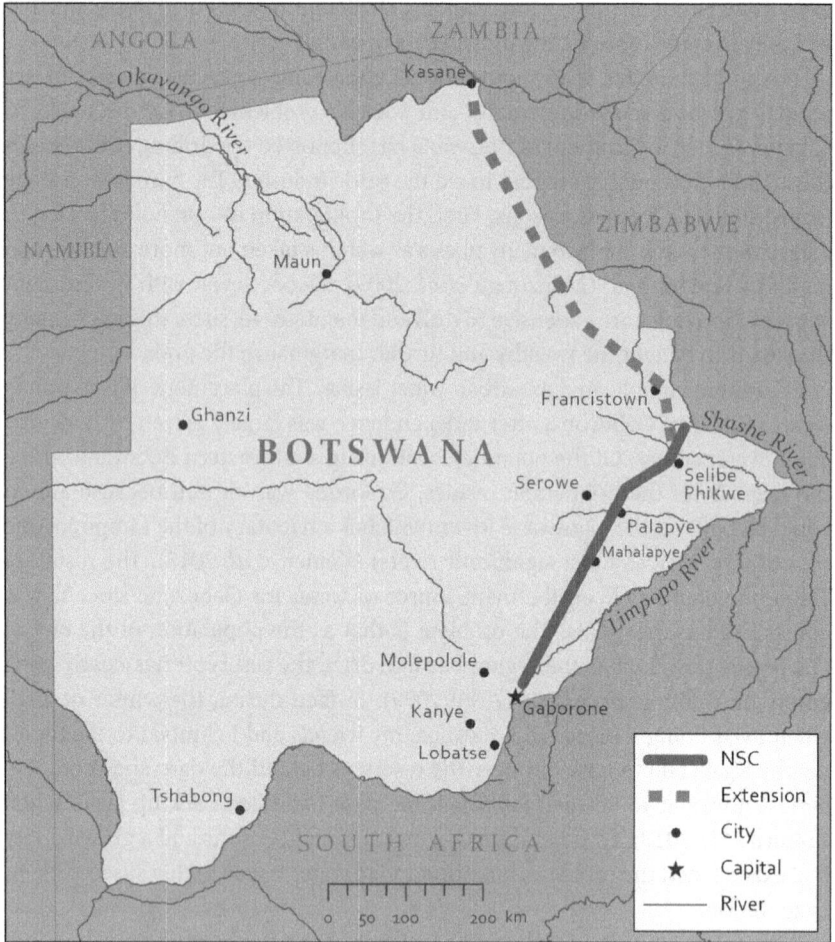

Figure 10.4 North–South water carrier (NSC) and potential extension
Cartography by Julia Castellano, Macalester College.
Sources: Simplemaps 2023; Esri Africa 2018, and World Agroforestry Centre 2014, UTM Zone 34S.

lead to a never-ending cycle of building pipelines (Moseley 2015c). A further problem with such water pipeline strategies is that large amounts of water are lost during the transfer and water diversion can have significant ecological and livelihood impacts on downstream aquatic life and human communities.

Of course, being semi-arid is not a new condition for Botswana and, as such, local people adopted strategies to conserve water. Foremost among these was the growing of drought tolerant crops and an extensive animal husbandry system that moved animals over large areas. As discussed in Chapter 6, many Botswana farmers have moved away from sorghum and towards maize because of labour constraints and a research enterprise that has favoured maize. Furthermore, the privatization of common grazing areas and extensive fencing (done to control the transmission of foot and mouth disease between wildlife and cattle) has restricted the movement of livestock.

While traditional and sustainable irrigation methods have been developed in other parts of the world, such as qanats in the Middle East,[5] Botswana likely does not have the right subsurface geology or population density to build and maintain these. That said, other methods hold great potential. The backyard gardening initiative was on the right track as it considered water conservation from the get-go. Most project beneficiaries received the following types of equipment: (1) rain barrels for rooftop water collection; (2) garden shade netting to reduce evapotranspiration; and (3) plastic tubing for drip irrigation. Drip irrigation is highly efficient form of irrigation originally developed by the Israelis (Van der Kooij *et al.* 2013). It basically employs an irrigation tube laid along a plant row that has small holes at the bases of plants letting out water. When combined with a water tank system, gravity will push the water out to the plants when a valve is opened. Shade netting, combined with mulching, also cuts down on evapotranspiration, which is quite high in Botswana's dry, sunny environment (especially in the warmer, summer months).

Another possibility that has yet to be extensively tested in Botswana is the use of check dams. Botswana's dryland landscape is crisscrossed by ephemeral sand rivers (also known as wadis in northeast Africa) that only run once or twice a season after a sudden rain event (common throughout the southern African drylands) (Walker *et al.* 2018; Mathias *et al.* 2021). The water from these events often quickly infiltrates into a sand aquifer where gardeners will access it by digging shallow wells. This process could be enhanced by constructing small gabion check dams to slow flow and increase infiltration into the sand aquifer. Check dams are a common water acquisition strategy in other dryland areas of the world such as Cyprus (Iacovides 2017). The benefits of this approach over traditional dam or water impoundment strategies number at least two. First, check dams are shallow structures that are mostly buried below the surface of a river bottom. When ephemeral rivers do flow, most of the surface water flows over the dam, thereby minimizing any downstream ecological consequences related to impeded river flow. Second, most of the water blocked by such dams

---

5. Qanats, also known as foggaras, are essentially underground aqueducts. Originally created sometime between the tenth to eighth centuries BC, these gravity-fed underground tunnels, which were common throughout the Middle East, would carry water from hillside underground aquifers to agricultural communities in fertile valleys. Their gravity-fed design meant that they did not overly deplete aquifers (Lightfoot 2000).

is stored underground in sand aquifers and is not amenable to evaporation like impounded surface water. As such, a key advantage of this strategy is that it minimizes water loss.

Of course, these strategies discussed regarding the agricultural sector will not address other municipal water uses in the City of Gaborone. For example, despite being located in a semi-arid area, the wealthy residents in the upscale Phakalane neighbourhood and golf estate maintain green, turf grass lawns, not to mention an 18-hole golf course (Kadibadiba *et al.* 2018). Such excessive water use is clearly unsustainable.

Conclusion

Set against an agricultural policy context that has largely benefitted male cattle owners, Botswana's backyard gardening initiative was one of the few agricultural initiatives in the country that mostly assisted women farmers. This initiative was subsequently panned as a failure and a waste of money. What these condemnations ignored were two critical structural problems with the programme (which actually worked quite well for the women involved when these impediments were addressed). The first problem had to do with the framing of the programme as a poverty alleviation effort that employed commercially oriented horticulture as the solution. By targeting households that fell below a certain poverty threshold, the programme often ignored the time poverty constraints faced by struggling households. Furthermore, entrepreneurial solutions aimed at the poorest of the poor often have a checkered track record as they do not acknowledge the situated (or structurally limited) agency of this population (Moseley 2014b). The second problem had to do with water constraints. Many of the women involved had to pay for expensive municipal water, which made it all but impossible to turn a profit on a small garden. In cases where women gardeners were not hobbled by these constraints, our research showed that women did quite well (Fehr & Moseley 2019). Curiously, the travails of this programme received a lot of attention in the Botswana media whereas the successes received little to no attention, suggesting – perhaps – that there is some level of gender bias in these discourses of failure.

Sustainably addressing water constraints in a semi-arid country afflicted by climate change is a major challenge. There are a variety of traditional strategies aimed at water conservation in Botswana (such as growing drought tolerant crops like sorghum), but there is also a fair amount of experimentation occurring in the water conservation and recuperation spheres that is not necessarily "traditional" in nature. This includes several approaches that were introduced alongside the backyard gardening initiative, such as water barrels, shade netting, mulching and drip irrigation. While agroecology is sometimes associated with traditional measures, it is important to keep in mind that farmers themselves have always experimented with new techniques. Furthermore, agroecology actively spans the experiential knowledge of farmers and ideas in the scientific community that are consistent with an agroecological approach (an approach that seeks to leverage ecological interactions in favour of food production, rather than strategies that work against nature). This complementary use of experiential knowledge and formal science is what makes agroecology a more decolonial approach. The combined use of both forms of knowledge has also long been argued for by certain scholars working in this realm (Richards 1985; Nyantakyi-Frimpong 2017). The key to unlocking a food secure future for Botswana lies in supporting women agriculturalists and innovating in the water conservation sphere.

11

AGRARIAN JUSTICE IN SOUTH AFRICA'S WESTERN CAPE

When I was doing fieldwork on land reform in South Africa's Western Cape Province, I had the pleasure of visiting land reform projects all over the province. The rural landscape of the Western Cape is dominated by large commercial farms, most of which continue to be run by white farmers. As discussed in Chapter 7, Europeans have been in this part of South Africa since the seventeenth century and they very quickly dispossessed most local people of their land in the early days of occupation. As such, this area was essentially the hearth of white farming in southern Africa, an area from which white farmers spread out into other regions of South Africa as well as neighbouring countries. This province never had homelands, which now operate as communal areas and are bastions for smallholder farming in other parts of South Africa. Given this, nearly all of the land redistribution projects I visited were large commercial farms that had been sold wholly or partly to farm workers, but were still run as commercial operations. Sadly, many of these farms were struggling.

Amidst this sea of large commercial farms, I had the opportunity to visit a former mission station. These communities are scattered across the Western Cape landscape and many of them were set up in the eighteenth century by various protestant missionaries. These stations had historically served as refuge areas for black and mixed-race South Africans, especially after slavery was abolished by the British in 1834. These island refuges (among a sea of white commercial farms) had also been home to some of the few smallholder farmers in the province. That day, I met the local agricultural extension agent. We walked all over the area, talking to residents, learning about the history of the town, and visiting a variety of gardens and small farm plots (see Figure 11.1).

Near the end of the day, I sat down with the extension agent to debrief on what we had seen. He began by apologizing that there really was no agriculture in the area anymore. This took me aback because we had just spent most of the afternoon wandering through plots of vegetables and chatting with farmers. What became increasingly clear to me over the course of this discussion with

Figure 11.1 Smallholder plots in (former mission station) Genadendal, South Africa
Source: Photo by author.

the agricultural extension agent was that he did not really consider small farms or gardening to be real agriculture. While I was excited to see so many people producing vegetables for themselves, their extended families and neighbours, and selling a bit on the side, for him it was child's play. It was as if everything we had seen that day was invisible to him. I do not blame the extension agent for this blind spot, he was very kind, generous with his time and clearly cared about his work. Rather I blame his training and the larger South African agricultural establishment, which has a particular vision about what does and does not constitute real agriculture (discussed in Chapter 7). Clearly, intellectual colonization runs deep in this part of South Africa, especially in the agricultural sciences, making the need for political agronomy and agroecology all the more apparent.

As discussed in Chapter 7, after the end of Apartheid, South Africa's African National Congress-led government adopted a market friendly approach to land redistribution that was heavily influenced by the World Bank's negotiated land reform model based on the principle of willing seller, willing buyer. Furthermore, the goal of land redistribution, especially after President Thabo Mbeki came to power in 1999, was to produce a class of black commercial farmers running large, production-oriented farms. These efforts have only led to incremental change and have been deeply frustrating for many members of South African society. This chapter explores alternatives to the conventional land reform approach in South Africa that emphasizes large-scale commercial agriculture. To this end, I explore two different alternatives. The first is a partial solution that emphasizes fair trade. The second is more promising and focuses on smaller farms that have a food security rationale rather than a business or commercial one. This is framed as an agrarian justice approach.

Is fair trade the answer?

One alternative to engaging with fiercely competitive global markets that squeeze as much as possible out of the farmers and workers that produce for them is to participate in markets that value a living wage and fair renumeration. The global fair trade movement represents one such an opportunity. In many cases, fair trade producers are also certified or de facto organic producers, suggesting that environmental and social sustainability often go hand in hand.

Amidst a larger research project examining land redistribution projects (Moseley 2007b), I studied a handful of worker-owned vineyards in the mid-2000s, some of which were in the process of becoming fair trade certified. As discussed in Chapter 7, the South African wine industry grew explosively after the end of Apartheid. Unlike some other agricultural goods where origin is less important, the global wine industry attaches great significance to the provenance of wines, with some of the world's most expensive wines coming from registered *terroirs* or territories (especially in Europe) (Meloni & Swinnen 2018). This attention to origin meant that a very limited amount of South African wine could be sold outside of the country during the Apartheid years due to international boycotts, especially in the 1970s and 1980s. This all changed after 1994 and South Africa quickly grew to being the eighth or nineth largest wine producer in the world. During the first ten years after the end of Apartheid, South African wine output grew from 38,850,664 litres in 1994 to 153,355,137 litres in 2004, a nearly quadrupling of production (SAWIS 2006). In this time period, South Africa also benefitted from a growing global awareness of Southern Hemisphere wines that were developing a reputation as good quality, yet affordable (Moseley 2008c). What was not clear was whether this phenomenal growth would benefit any of South Africa's historically disadvantaged groups.

In addition to land redistribution, the ANC was also interested in increasing the participation of Black South Africans in agri-business decision-making (and across the private sector), an approach known in South Africa as Black Economic Empowerment (BEE) or even more broadly as transformation. As a result of transformation policies, altruism, or the prospect of business opportunities, some white vineyard owners had voluntarily gone into partnership with their black workers. Others had sold all or part of their vineyards to black entrepreneurs, yet essentially left the working conditions and the structure of decision-making unchanged (Williams 2005). Still other white farmers sought to improve labour practices on their vineyards, yet maintained majority ownership. Finally, some obtained fair trade certification. These vineyards often marketed their wine as worker-owned, black-owned or fair trade certified. As a consequence, South Africa's Western Cape was dotted with a small number of vineyards that had alternative ownership or labour arrangements. As mentioned previously, in the midst of a broader study of land reform efforts, we conducted a narrower study of vineyards to see if any of these of alternative arrangements were leading to qualitative improvements for the black farm workers involved (Moseley 2008c).

Wine produced on farms that were worker-owned (because of land redistribution programmes) or had BEE status may have meant something to South African consumers, but support for these wines within South Africa was limited (Moseley 2007c). Furthermore, these distinctions meant little to the broader, international wine market (Moseley 2006b). While fair trade certification came to the wine industry later than other products such as chocolate, tea and coffee, South Africa was at the forefront of this process globally with the first two certified fair trade wine producers in 2003 (FLO 2008a). Very interestingly, South African labour activists and academics then opted to develop South Africa specific fair trade standards in 2004 that were more demanding than the global standards. As Kruger and Du Toit (2007) describe, there was concern that extending the standard fair trade criteria to South Africa unchanged would undermine local motivations for transformation. The typical fair trade labour requirements for farm workers on large, commercial farms (or plantations) called for decent wages, good housing, meeting health, safety and environmental standards, and ensuring the right to join unions. While these criteria may have made sense internationally, they were no stronger than the new protections that the ANC had put in place in the post-Apartheid era. The last thing labour activists wanted was to market South African wine as socially conscious, especially given the long history of labour exploitation discussed in Chapter 7, without additional requirements attached to fair trade certification. The compromise, after much negotiation with international certifiers such as Germany-based Fairtrade International (FLO), was that farm workers needed to own at least 25 per cent of the farm, and they needed to be involved in management decisions, in order to be fair trade certified in South Africa, a stricter standard than the international one (Kruger & Du Toit 2007; FLO 2008b).

There was a mixed set of results for the worker-owned and fair trade vineyards examined in this study, from those doing extremely well, to others who had outright been taken advantage of by a white co-owner (Moseley 2008c).

Herewith a few of the more positive examples. Bouwland vineyard came into being in 2003 when 60 farm workers (split between men and women) benefit-ted from a government land redistribution grant. They purchased a 76 per cent share of a 56-hectare vineyard as well as the associated Bouwland wine label (which is important because most of the money is made selling the wine, rather than the grapes). They have 40 hectares in Pinotage, Cabernet Sauvignon and Merlot grapes. Their white partner (who has a minority share in the business) is a well-respected winemaker for nearby Beyerskloof and Kanonkop vineyards. The property has an old farm house that it uses as an office and tasting room. The group relies on Beyerskloof Vineyard for equipment and on co-owner Byers Trutor for his wine-making expertise. While a couple of the worker co-owners work full time for Bouwland, the majority have kept their day jobs working at Beyerskloof and Kanonkop vineyards. The worker co-owners pay themselves to work on the Bouwland vineyard in a regular rotation of shifts between the three farms. Bouwland is turning a modest profit (which would be larger if they were not paying off a loan which supplemented the government grants to buy the farm). The group is producing and selling 17,000 cases of wine per year that is sold locally and exported to the Netherlands, Germany, the UK and the US. The group markets their wine as worker owned, but it is not fair trade certified (Moseley 2008c; Clarke 2020). This farm has promise and they were fortunate to have a well-intentioned co-owner and collaborator, but some of the workers worry about their dependence on their white co-owner.

The Nuutbegin Trust (which is Afrikaans for new beginning or fresh start[1]) was initiated in 2000 when workers from nearby wineries Waterkloof and Fransmanskloof vineyards obtained a government land redistribution grant to purchase a 50 per cent share in a long-term lease on 25 hectares of prime vineyard land. The other two partners are the white owners of Waterkloof and Fransmanskloof vineyards who each have a 25 per cent share in the ven-ture. Nuutbegin, along with two other vineyards, produce Merlot, Shiraz and Cabernet Sauvignon grapes for the fair trade certified Thandi winery (which Nuutbegin has a 7 per cent stake in). Thandi was the first winery in the world to obtain fair trade certification (Bek *et al.* 2007). Like Bouwland, most of the worker co-owners have maintained their day jobs on nearby wineries and they

---

1. The reader may be puzzled as to why a community of coloured workers would use a name in Afrikaans (the language of the oppressor). The reality is that this is the first language of this mixed-race community that no longer speaks the indigenous languages of their forebears (not unlike other formerly enslaved peoples around the world). In fact, there is an active effort by some in the coloured community to claim the Afrikaans language as one that they actively co-developed alongside Dutch European settlers. In other words, Afrikaans is an older form of Dutch that has lots of African linguistic influences that have not traditionally been acknowledged in White South Africans' narratives of their linguistic history (Roberge 2002).

coordinate their work schedules to spend time working on the Nuutbegin vineyard. The big difference from Bouwland is that they are fair trade certified, which was a major expense to acquire. It remains to be seen if this investment will pay off over the long run (Moseley 2008c).

South Africa-specific fair trade standards hold some promise because they call for at least partial worker ownership and offer distinction and a higher premium in the market place. Nonetheless, there remain a number of limitations with this model. First, it is expensive to become fair trade certified, and many wineries cannot afford this, especially small ones. Second, workers owning or co-owning a winery is challenging if they do not have the management training, experience and contacts to be successful. This can make it difficult for them to survive on their own or may put them in an unequal position with a white co-owner who has such skills and experience. Third, there are limits to the fair trade model for producing social change. While informed and ethical consumer choices can aid change, it cannot substitute for needed government investment in education and legal support for black farmers when there is malfeasance. Lastly, fair trade does not necessarily promote a new model of production but operates within the existing system of large, commercially oriented farms (Moseley 2008c). In sum, this approach may be limited in its ability to address underlying structural problems.

Land reform that privileges small-scale farming and food security

As discussed in the introduction to this chapter, the former mission stations of the Western Cape are some of the few places where smallholder farming has a history and continues to persist as a tradition. While the agriculture in these areas is essentially invisible to the provincial agricultural authorities, my former student Michaela Palchick and I sought to understand the food security and livelihood benefits of these activities (Palchick 2008; McCusker *et al.* 2016a). With a different student, Megan Grinde, we studied one of the few cases in the Western Cape Province where large commercial farms were purchased with land redistribution grants and then broken up into smaller holdings to be managed by individual farmers (Moseley 2007b; Grinde 2008). Both of these examples of smallholder agriculture are explored in this section.

Smallholder farming on former mission stations

The Act 9 areas or "coloured reserves" were an Apartheid category of land in the Western and Northern Cape Provinces. In many cases, these areas were originally mission stations set up by various protestant groups in the eighteenth and nineteenth centuries, but then subsequently nationalized and run by the state (often as a de facto labour reserve for the surrounding white farming area). As noted previously, these areas served as refuges for the Khoisan population and other groups after enslaved people were freed in the 1830s. These were areas where small-scale agriculture and animal husbandry persisted and evolved because of people's marginal status.

There were 23 Act 9 areas that were designated for coloured occupation during Apartheid, covering 1.8 million hectares in four provinces (Catling 1996). Collectively, these areas were one tenth the size of the formal bantustans and contained much smaller populations. While similar in some ways to the bantustans, they were also different in important ways. First, these were not the historical homelands of the resident population, but rather people moved there. Second, the subsistence traditions in Act 9 areas were not what we might typically think of as indigenous. Rather, they are a product of external constraints because the regime did not allow people of colour to be commercial farmers. The farming methods these farmers employed were a result of different influences, namely farming approaches promoted by the missionaries, the methods of white commercial farmers in surrounding areas, and various traditional practices within the coloured community. Third, most Act 9 areas were relatively small and surrounded by white farming areas so there were always connections to outside areas in terms of labour flows and ideas (McCusker *et al.* 2016a).

The rest of this section is devoted to a particular case study of the community of Genadendal, which I briefly discussed in the introduction. My student Michaela Palchick lived here while doing research, and I had the good fortune to visit multiple times. The community was established in 1738 by German Moravian missionaries. It is located 100 kilometres to the northeast of Cape Town in the foothills of the Riviersroderend Mountains in the Overberg District (see Figure 7.3 in previous chapter for map). At one point, Genadendal was briefly the second largest town in South Africa. The land is no longer controlled by the Moravian Church but held in trust by the central government for the local people. Most people in the town identify as descendants of the Khoi Khoi, the German Moravian missionaries, or people freed of enslavement. Anyone who can claim familial ties to the community has the right to a plot of land which is managed as a commons[2] with usufruct rights held by different families (which is quite different from the private property regime in surrounding white farming areas) (Palchick 2008). The community has 1,250 households and a total of 4,000 hectares of land. The biggest village is Genandedal, but there are also three smaller ones, namely Bereaville, Voorsterkraal and Boschmanskloof.

---

2. Common property regimes were originally discussed in Chapter 1.

Our research detected 35 medium-scale, semi-commercial producers as well as over 100 gardeners. The first group was mostly male and they had an average plot size of five hectares. Table 11.1 shows the average age and gender distribution for the gardeners in our survey, which skewed older and male. The average plot size was 700 m². Most of our interviewees had learned to garden from their parents or grandparents. The main crops they grew included potatoes, sweet potatoes, pumpkins, butternut squash, tomatoes, peas, green beans, onions, cabbage, peppers, carrots, Swiss chard, lettuce, makatons (South African melons) and cucumbers. Most of the gardeners produce enough to supply themselves and their extended families with vegetables during the growing season.

To give the reader a better sense of gardening livelihoods in Genadendal, here I describe two gardeners in slightly greater detail. The first gardener was 33 years old, and his day job was working for the government water conservation programme. He worked in his gardens in the evenings and on the weekend. He had been gardening for eight years and learned to do this from his parents. He had three plots. One was his own, and the other two he borrowed from his neighbours in exchange for a share of the vegetable crop. Sometimes his children helped him to work in the gardens. He grew potatoes, green beans, pumpkins, onions, gem squash and tomatoes. He typically would sell 50 per cent of his crop, and the other half was for home consumption. He sold his produce locally and in the nearby town of Greyton. What he grew covered most of his family's vegetable and tuber needs, except for a few months in the winter. He also had two pigs, a horse and a milk cow. He typically planted twice per year in August–September and then again in March. For context, summer runs December–March and winter from June–August, so most of his vegetable production was in the spring and autumn. Some seeds he saved from season to season, whereas others he bought. He rotated his crops to minimize pest problems and used manure from his animals to fertilize the fields. In sum, the garden cost him little in cash outlays to run, but it provided him and his family with a valuable supplemental income and a very important source of food.

The second gardener was 43 years old and she worked for the municipal government. She had two 40-by-20 metre plots (see Figure 11.2). One plot was hers, and the other she borrowed from an elder who could no longer garden. She

Table 11.1 Survey of gardeners in Genadendal

	Age	<40 Years	40–60 Years	>60 Years	Total
	Female	3	8	22	33
Gender	Male	11	17	42	70
	Total	14	25	64	103

Source: Palchick (2008).

Figure 11.2 Smallholder farmer on her plot in Genadendal, South Africa
Source: Photo by author.

learned how to garden from her mother and grandmother. In the off season, she let cows graze in the gardens to eat the stubble and manure the fields. She hired someone at the start of the season to plough the two plots in September. She then took some time off from her day job to plant, but then maintained the gardens in the evenings and on the weekends. She then typically harvested in December and then planted again in January, harvesting again in March. She grew sweet potatoes, potatoes and a variety of vegetables. According to her calculations, she consumed about 20 per cent of the garden produce herself, shared another 40 per cent with her extended family, and then sold about 40 per cent locally. She reported that she earned a net profit of about 2,000 rand a year from selling her vegetables at the local market (over the course of two three-month seasons). For context, at the time we did this research the average monthly salary for a farm worker was 450 rand, so she was earning the equivalent of about 4.4 months in wages.

While agriculture in Genadendal is somewhat invisible to the conventional agricultural establishment in South Africa, because it is not considered real farming, what the experience of gardeners here suggests is that it can have real food security benefits. These are not big commercial farms but small intensively cultivated plots that produce enough vegetables and tubers for the family, as

well as extra to sell at the market. The health, food security and income benefits of farming on the side should not be trivialized, especially since large-scale,
commercially oriented farming has often ended in bankruptcy for the few land
reform beneficiaries who were lucky enough to get grants in the first place. Of
course, the huge advantage that residents of Genadendal and other Act 9 areas
hold is secure and debt free land tenure. Could this option be made more widely
available? A small, but debt free, land holding producing food for home consumption is a powerful yet overlooked tool for enhanced food security.

The smallholder farming approach to land redistribution

Although the vast majority of land redistribution initiatives in South Africa have
been based on the large commercial farming model, wherein relatively large
groups of beneficiaries collectively run commercial farms as an enterprise, there
have been a handful of isolated cases where bigger farms were split up into
smaller parcels and then managed as individual smallholder plots. My former
student Megan Grinde and I found such an example in 2007 and it was the subject of Megan's thesis research (Grinde 2008). While this case was by no means
a clear success, it is worthy of discussion as an alternative.

According to an oral history of the Bokdrif land redistribution project assembled by Grinde (2008), the idea to develop a smallholder approach in this case
emerged from discussions between a local agricultural extension agent and
one of the eventual beneficiaries (who goes by the pseudonym Georg) of the
project. Both had noticed that projects with a large-scale commercial orientation in their area were not doing well. As such, they began developing a smallholder approach in their application for a Land Redistribution for Agricultural
Development (LRAD) Grant. Georg was well networked with other interested
farmers via training he had done with a non-profit called the Atlantis Small
Farmers Association (AFSA) where he had learned about food production techniques. The group Georg assembled consisted of eight families and 35 people.
They wished to purchase a 31-hectare parcel that would be split into eight family
plots of a little less than four hectares each. This land was a section of a larger farm
that a local white farmer wished to sell because it was less productive (which,
as discussed in Chapter 7, was a common problem with South Africa's market-
based approach to land reform). Their application initially received a considerable amount of push back from the Department of Land Affairs because it was
not commercially oriented enough. Land Affairs did not like the idea of breaking
up the land into smaller plots or the focus on food production for household
consumption. But the group persisted and was eventually successful by making
the argument that they would gradually transition to a more commercial mode.

The other thing the Department of Land Affairs wanted was for the group to take out a loan from the local land bank to allow for a stronger commercial orientation. But the group was wary of loans because they had seen many other projects fail when they defaulted. Again, the group persisted, insisting that they did not want to take this risk. They eventually prevailed and received a $100,000 grant to purchase the 31 hectares that was split between the eight families. They bought the land outright with no debt.

The group encountered additional bureaucratic difficulties getting water and electricity delivered to the site. It also took a while for them to get permission to build an access road off of the main highway into their farms. Eventually five of the eight families moved to the site and began farming (see Table 11.2). Most of them grew vegetables and raised small stock. While none of the families were self-sufficient, as in Genandendal, these small farms were making important contributions to their household food security. Some were also planting perennial tree crops (such as olives) that would bear fruit over the longer term. They also were able to sell some of their produce locally, although the supermarket chains in the larger towns did not like dealing with smaller quantities of produce. Perhaps most importantly, these farmers felt agency in growing their own food and owning their own land, something Apartheid had taken away from their people for the last 350 years. According to Grinde (2008: 80), when she interviewed one of the families, "Helen speaks in glowing terms of farm life, Georg nearly bursts with excitement when the subject comes up ... He is deeply proud of his achievement in this farm of his own".

Conclusion

While the large-scale commercial farming model may be appropriate for some land redistribution beneficiaries, South Africa's land reform initiatives must consider supporting other agricultural models, such as agroecological and smaller scale commercial farming. Given the history and dominance of large-scale commercial agriculture in the Western Cape Province of South Africa, and the virtual extirpation of small-scale commercial or subsistence farming in the area, a bias towards the former is not all that surprising. However, given the incredibly competitive nature of the global agricultural products market, the extensive capital and contacts needed to thrive in this arena, and an emerging set of environmental issues associated with large-scale commercial agriculture (such as pesticide resistance), it does not make sense for the government to push all emerging farmers into this domain.

Fair trade offers some promise as an alternative model because it offers a niche market with higher premiums, and it values fair wages and good working

Table 11.2 Division and use of plots at Bokdrif

Beneficiary household*	#	Resident?	Plot manager*	% of plot in use	Plot use	Farming as % of income	Other income	Cultivars	Livestock
Seppie	1	Yes	Seppie	90%	Vegetable cultivation, limited livestock forage	20%	Manages his own electrical repairs business	Cabbages, onions, green beans, yarrows, pepperdews, sweet melon	pigs, chickens
Jansen	2	Yes	Jansen	75%	Vegetable cultivation, limited livestock forage	20%	Owns multiple mini-taxis	Cabbages, onions, green beans, yarrows, pepperdews, sweet melon	pigs, chickens
Alexander	3	No	Daniel, grandson of beneficiary	75%	Vegetable cultivation, limited livestock forage	0%	Works as dentist assistant in Malmesbury	Cabbages, onions, green beans, yarrows, pepperdews, sweet melon	pigs, chickens
Georg	4	Yes	Georg	80%	Vegetable cultivation, limited livestock forage	20%	Disability pension, lets house in Malmesbury	Cabbages, Onions, Olives, green beans, yarrows, pepperdews, sweet melon	pigs, chickens
de Water, Sr.	5	Yes	de Water, Sr.	100%	Livestock forage	25%	Railway pension	none	goats, chickens
Frans	6	No	de Water, Sr.	100%	Livestock forage	0%	Employed in Malmesbury	none	none
Appelgrein	7	No	Seim, friend from Malmesbury	75%	Vegetable cultivation, limited livestock forage	0%	Farm in the Northern Cape	Cabbages, onion, limited vegetables for consumption	chickens
de Water, Jr.	8	Yes	de Water, Jr.	20%	Limited vegetable cultivation	10%	Employed in Malmesbury	Limited vegetables for consumption	chickens

*Names have been changed to protect the anonymity of the beneficiaries
Source: Grinde (2008).

conditions. As such, it avoids a race to the bottom that we see in other capitalist markets. Nonetheless, even fair trade farms for products like wine tend to still use the large commercial farming paradigm. While this model may work for some, it requires extensive commercial contacts and significant capital investment, which make it a risky endeavour.

In many instances it may be more reasonable for emerging farmers to pursue a model of small-scale commercial or agroecological farming. In practical terms, this means allowing large commercial farms to be divided into smaller plots to be managed by small groups, individuals or households. While this approach was permitted under the original land redistribution programme (known as SLAG), it has been virtually non-existent since 2000. While some policy-makers expressed concern about a lack of markets for small-scale producers, interviews with smallholder farmers in the Western Cape revealed that these producers were not at a loss for such markets. Informal markets for vegetables, fruits and small stock abound in the province's black and coloured settlements. While animal traction and the use of organic inputs may be viewed as backward by the province's influential big farming interests, the fact is that farm operations employing this model have lower input costs, lower debt burdens, and fewer environmental problems. While experimentation with this model in South Africa has been quite limited, the experience of smallholder farmers in the former coloured reserves suggests that it can lead to substantial food security improvements. Furthermore, in the very few cases where land redistribution projects did allow large farms to be broken up, the results appear to be quite promising in food security terms. Smallholder farmers also feel a great deal of pride and agency in what they are doing.

In sum, transformation in the Western Cape's agricultural sector should be as much about redefining the dominant farming model as it is about involving people of colour in this enterprise. A more varied farmscape in the Western Cape Province, involving large-scale commercial and smallholder farmers, may actually mean that formal economic measures decline as non-market transfers of food increase. The feared collapse of commercial farming in South Africa has been the bogey man that the commercial farming lobby has used to beat back any efforts to experiment with a smallholder farming focused land reform approach. But the big commercial farming approach is not working. Food security measures are not improving and most people in the black and coloured community feel no sense of justice. It is time to have a more differentiated approach that includes the promotion and support of smallholder farming. A more varied farmscape in the Western Cape (with small and large farms) could lead to less poverty, more autonomy, fewer environmental problems and enhanced food security, that is, greater agrarian justice (Moseley 2007b).

PART IV

AFRICAN AGRICULTURE AND GLOBAL POLITICAL ECONOMY

.

12

FEEDING OR FIGHTING THE FIRE: INTERROGATING THE INSTITUTIONAL ARCHITECTURE SHAPING AFRICAN AGRICULTURE, FOOD SECURITY AND AGROECOLOGY

In 2016 I was tasked by an international agricultural research centre to do a political economy analysis for a crop-livestock integration project in Burundi. At the time, Burundi (a small land locked country in East Africa) was an isolated country and the ruling regime was considered to be deeply problematic given its autocratic tendencies and human rights abuses (UNHCR 2017). Given its international pariah status, many other development actors and donors had left the country, leaving this Collaborative Group for International Agricultural Research (CGIAR) centre as one of the few remaining donors and development actors. Burundi is one of the poorest countries in the region and the world, ranking 187 out of 191 in terms of the 2021 Human Development Index (UNDP 2022). It is also one of the most densely populated countries in Africa and soil degradation is a growing problem. While conventional agricultural interventions might emphasize improved livestock breeds, certified seeds and external inputs like inorganic fertilizers, this seemed highly impractical in a country that was both impoverished and deeply isolated with a limited ability to trade. In the face of such constraints, what the local staff had developed was a deeply innovative project that involved local livestock breeds, animal penning and low external input agriculture (Moseley 2022c).

More specifically, the main idea of this crop–livestock integration project was to: introduce penning (rather than free-range care) of animals; collect manure from the penned animals, which was composted and spread on fields; and then use the crop refuse, as well as some intercropped forage species, to feed the animals. The project worked with farmers of varying income levels to experiment with several locally appropriate crop–livestock combinations, including: (1) cattle with maize; (2) pigs with cassava, sweet potatoes and soybeans; (3) chickens with beans; and (4) rabbits with vegetables. The idea was that households have different capacities to care for animals, so better off households might be introduced to combinations 1 and 2, whereas households with less capacity might adopt combinations 3 and 4. This intensive, and integrated, crop-livestock approach was ideal for an isolated country with limited ability to trade and in which farm expansion was seriously constrained, soil fertility was declining, most rural people lacked sufficient protein and many farmers had limited means to purchase commercial inputs. While not framed as such, in working with the realities of Burundian farmers, the local staff had developed an agroecological approach that allowed for intensification (Moseley 2022c).

My draft report summarized these political economy constraints and suggested that the approach that had been developed by the local staff made a lot of sense in this context. I further suggested this model could serve as an alternative to the value chain model and New Green Revolution for Africa approach that most international agricultural organizations were pushing in Africa. This draft report then went to higher ups in Addis Ababa for review and – to make a long story short – the response was mixed. They appreciated the analysis that was undertaken, but they were deeply unhappy with my open critique of the value chain model and the New Green Revolution for Africa approach, which they considered to be heretical (my language not theirs). While I did not retract my assertions from the final report, I did soften my language and was more diplomatic so that it could be accepted.

This experience raises at least three important points in relation to this chapter and the broader themes of this book. First, dominant narratives and the status quo approach get perpetuated over time by seemingly small acts of confirmation and reinforcement. Leach and Mearns (1996) discuss how development consultants are one such set of actors. Most consultants are under extreme time pressure to produce a report and, perhaps more importantly, they often do not want to ruffle the feathers of those who are employing them because they would like another contract. Both factors make it likely that many consultants will find it easier to confirm the status quo thinking. I had more agency to push back in this situation as I was already employed as a professor and thus did not depend on this type of income for my livelihood. If I am honest with myself, I can imagine behaving differently if I were living contract to contract. Second, understanding knowledge politics is central to political agronomy (Andersson & Sumberg 2016). As Vanloquerin and Baret (2009) write, an analysis of knowledge politics in agronomy, and associated development organizations, helps explain why particular technologies or development pathways are privileged over others. CGIAR centres have a role to play in this process as they can either reify a productionist understanding of food problems and solutions through their hierarchies, or point the way towards different approaches. Finally, alternative agricultural development and food security paradigms such as agroecology need support from higher levels if they are to grow and thrive. This includes governance, funding, research and education.

There is a common refrain in the environmental community that calls on people to "Think Globally and Act Locally".[1] As I have repeatedly hinted at in various stages in this book, and should be abundantly clear from the opening example to this chapter, the problem is that local action is conditioned by broader-scale policies and programmes. As such, acting locally often necessitates pushing back globally if real change is to occur. The rest of this chapter will examine the regional and international context shaping African agriculture and food systems. To build more resilient food systems and a different kind of agricultural development, African farmers and leaders will need to navigate the international and regional institutional architecture that built and sustains the existing global food system, such as the CGIAR centres and the UN Food and Agriculture Organization (FAO), as well as regional organizations like the African Union. To really take hold and transform African food systems, agroecology and a six-dimensional food security approach need to be supported at multiple scales and higher governance levels. As such, this chapter highlights a case study of a science–policy–society interface where this seems to happening, a situation that is not unrelated to its more participatory and inclusive structure. The chapter ends by briefly discussing a regressive counter insurgency that has attempted to marginalize an insurgent agroecology and reassert the prominence of the production agriculture paradigm.

The international and regional institutional architecture supporting the status quo

As discussed above, the way people interact with and manage their environment does not exist in a vacuum, but rather is shaped by local dynamics as well as policies and programmes at the national, regional and international levels. In other words, while local people have agency to shape their own futures, this agency is situated or conditioned by broader scale political economy. This multi-scaler perspective is a basic insight of political ecology. As such, postcolonial, productionist agricultural approaches to addressing hunger in the African context, from the first Green Revolution, to agricultural trade, to the New Green Revolution for Africa, have not just been pushed by agricultural agents at the local level but supported by funding mechanisms, commercial interests and research institutions at the national, regional and international levels.

---

1. The term "think globally, act locally" is often originally attributed to Patrick Geddes, an early twentieth-century Scottish town planner. Others say it was coined by David Brower in 1971 when he founded the organization Friends of the Earth (Wikipedia 2023).

The Green Revolution approach and its subsequent iterations did not just spring into being on its own. The translation of these ideas from basic research to implementation was initially supported by powerful American foundations, most notably and Rockefeller Foundation and the Ford Foundation (Schurman 2018). But then, perhaps most importantly, it became engrained in public institutions. On the implementation side, the FAO played a leading role in aiding governments in implementing Green Revolution practices (Patel 2013). In most cases, the FAO had technical experts or advisors embedded in the ministries of agriculture of various governments around the world. These advisors often had UN-funded agricultural projects to implement, but they also provided advice more broadly that would shape the agricultural policies of any given country. For example, the FAO actively supported the development of irrigated rice production in The Gambia in the 1970s (Carney 1993).

Collaborative Group for International Agricultural Research (CGIAR)

The other big area of institutional investment has been in the area of research, both basic and more applied research. The most notable international institutions in this regard are a network of UN sponsored research institutes known as the Collaborative Group for International Agricultural Research or CGIAR (one of which was discussed in the introduction). These research centres emerged in the post-Second World War period as a part of UN research and development efforts buttressing the Green Revolution (Bebbington & Carney 1990). CGIAR has an international coordination hub in Montpellier, France, but then there are 14 CGIAR institutions, focused on particular crops or themes, spread all around the world (see Table 12.1). Not only do these institutions undertake research at field stations, conducting field trials with new crop varieties for example, but they also have research action projects where they work closely with communities.

While CGIAR institutions receive some funding from the CGIAR trust fund, they also increasingly rely on grant funding from bilateral donors and private foundations. After the trust fund, the top five donors in 2022 were the Bill & Melinda Gates Foundation, the United States, the World Bank, Germany, India and the European Union (CGIAR 2023b). Given the predilections of donors, the CGIAR organizations have also tended to shift their priorities over time according to agricultural development trends, from the initial Green Revolution, to the focus on trade and comparative advantage, to a value chain and New Green Revolution for Africa emphasis.

The CGIAR system did launch an agroecology initiative in 2022, with the Center for International Forestry Research (CIFOR) and World Agroforestry

Table 12.1 The Consortium of International Agricultural Research Centers (CGIAR) and its organizations

	Centre Name	Headquarters Location
1	Africa Rice Center	Abidjan, Côte d'Ivoire
2	Center for International Forestry Research (CIFOR)	Bogor, Indonesia
3	International Maize and Wheat Improvement Center (CIMMYT)	Texcoco, Mexico
4	International Center for Agricultural Research in the Dry Areas (ICARDA)	Beirut, Lebanon
5	International Crops Research Institute for the Semi-Arid Tropics (ICRISAT)	Patancheru, India
6	International Food Policy Research Institute (IFPRI)	Washington, DC, USA
7	International Institute for Tropical Agriculture (IITA)	Ibadan, Nigeria
8	International Livestock Research Institute (ILRI)	Addis Ababa, Ethiopia and Nairobi, Kenya
9	International Potato Center (CIP)	Lima, Peru
10	International Rice Research Institute (IRRI)	Los Baños, Philippines
11	International Water Management Institute (IWMI)	Colombo, Sri Lanka
12	The Alliance for Bioversity International and International Center for Tropical Agriculture (CIAT)	Cali, Columbia and Rome, Italy
13	World Agroforestry (ICRAF)	Nairobi, Kenya
14	WorldFish	Penang, Malaysia
15	CGIAR system headquarters	Montpellier, France

Source: Compiled by author based on information from CGIAR (2023a).

(ICRAF) being the lead centres, along with bilateral partners Biovision, GIZ, CIRAD (which are, respectively, Swiss, German and French research and development organizations) (CGIAR 2022). The effort involves establishing Agroecological Living Labs in seven countries, including four that are in African countries: Burkina Faso, Kenya, Tunisia and Zimbabwe. This is a laudable effort, but inevitably this may lead to a clash of science cultures if it has not already. While the initiative is still young, there are some early signs that give pause for concern. Ideally, agroecology would be given institutional space within the CGIAR system to flourish by having, for example, a centre of its own. Absent this, agroecology could become marginalized (and its custody under some less powerful CGIAR centres hints at this possibility) or coopted by more mainstream agronomic thinking. Already, some CGIAR documents discuss how this approach could be scaled up (which some consider a contradiction with a place-based approach that emphasizes local knowledge (Moseley 2017b) or meshed with business models (CGIAR 2022)).

The African Union (AU) and Comprehensive Africa Agriculture Development Programme (CAADP)

At the regional level on the African continent, the African Union is arguably one of the most significant actors. The African Union is also, in some ways, the neoliberal reincarnation of the Organization of African Unity (OAU), which was a pan-Africanist and anticolonial body formed in the era of African independence (1963 more precisely). Over the course of the 1980s, the OAU increasingly fell into disrepute as it came to be known as a good old boys' club that turned a blind eye to coups and other violations of democratic norms. The OAU was disbanded in 2002 under the leadership of its last president, then South African president Thabo Mbeki. Mbeki then simultaneously launched the African Union (AU) in 2002 as its replacement.

The AU is supposed to be more inclusive and democratic than its predecessor and have more of a development mandate. Many of the AU's development initiatives were spearheaded under the New Partnership for Africa's Development (NEPAD), including the Comprehensive Africa Agriculture Development Programme (CAADP) which was launched with the help of the FAO (Munro & Schurman 2022). According to NEPAD, CAADP is "Africa's policy framework for agriculture and agriculture-led development". Furthermore, NEPAD argues that "agriculture and the food industry can be the engine for growth in Africa's largely agrarian economies, with tangible and sustainable impact on improving food security and nutrition, contributing to wealth and job creation, empowering women and enabling the expansion of exports" (NEPAD/CAADP 2023: 1). The framing of CAADP was consistent with a long line of previous initiatives that saw the development of agriculture as the first step in the economic growth process (discussed in Chapter 1), while also giving a nod to food security and women's empowerment. In the aftermath of the global food crisis of 2007–08, NEPAD and CAADP formed Grow Africa, an initiative to facilitate "collaborations between governments, private companies, and smallholders that would both lower the risk and cost of investing in African agriculture and mobilize investment into priority value chains" (Munro & Schurman 2022: 33–4). This is also the time period when the Gates Foundation-supported Alliance for a Green Revolution in Africa (AGRA) emerged (discussed in Chapters 2 and 5), an entity that began working hand and glove with Grow Africa.

What should be clear is that African institutions supporting agricultural development on the continent had become thoroughly captured by the neo-productionist, market-oriented agricultural development paradigm. With ample resources to fund projects, this thinking then reinforced attitudes within agricultural ministries and schools of agriculture across the continent, many of which were staffed by civil servants and professors who already leaned in that direction because of their training in development agronomy. This has largely meant that agroecology as an alternative paradigm has been left in the hands of farmers and civil society organizations such as the Alliance for Food Sovereignty in Africa.

An opening for agroecology in international science–policy–society circles

Science–Policy Interfaces (SPIs), which are sometimes also referred to as Science–Policy–Society Interfaces (SPSIs) depending on their structure, have been set up as bridges between the science community and the policymaking community on different topics. The existence of these types of structures is premised on the belief that good policy is based on evidence and an up-to-date scientific understanding of problems. For example, we would not want public health policies to be based on outdated understandings of disease dynamics. In most cases, SPIs have scientific panels composed of leading researchers that synthesize the best known information on a given topic as well as related policy recommendations. In some, but not all cases, these recommendations then get debated by policy-makers and turned into voluntary guidelines or binding international agreements. Perhaps the best known example of an international SPI is the International Panel on Climate Change (IPCC). This panel of international climate scientists regularly meets to discuss climate change and periodically publishes comprehensive assessment reports (written by teams of hundreds of scientists working on a pro bono basis) synthesizing the most up-to-date scientific understandings of the biophysical and human dimensions of climate change. To date, the IPCC has published six such reports, with the latest one being released in 2023 (IPCC 2023). These reports informed an international agreement, United Nations Framework Convention on Climate Change, as well as annual conference of party (COP) meetings that gauge progress against this framework.

While less well known, there are also SPIs that deal with food security and nutrition. Perhaps the most significant of these is the High Level Panel of Experts for Food Security and Nutrition (HLPE-FSN), which informs the work of the Committee on World Food Security (CFS). The CFS was created in 1974 to coordinate policy between the three UN Rome-Based Agencies (RBAs): the World Food Programme (WFP), the FAO and the International Fund for Agricultural Development (IFAD). In 2007–08, the CFS was caught rather flat footed with regard to the 2007–08 global food crisis. As a result, several reforms were made in 2009 that made it more inclusive, participatory and science-based (Eklin *et al.* 2014). Commentators like Olivier De Schutter have argued that these changes were "perhaps the single most significant development in the area of global food security in recent years" (De Schutter 2014). At least two major changes were made as a result of these reforms. First, a new institutional structure was created that included a broader range of stakeholders. Now instead of just representatives from member states, there was a private sector mechanism, a civil society mechanism (now the civil society and indigenous peoples mechanism), philanthropic foundations, financial institutions, research institutions, and UN organizations that work on food and agriculture. Second, a new body was created to provide scientific advice and reports to the CFS called the High Level Panel of Experts for Food Security and Nutrition (HLPE-FSN). The HLPE-FSN has a steering committee of 15 scientists that oversee the production of reports (compiled by teams of scholars) that are done at the request of the CFS. These reports go out for peer review by experts as well as public comment before they are finalized. See Figure 12.1 for a diagram of the organizational structure of the CFS and its relationship to the HLPE-FSN. In the interest of transparency, I note that I served two two-year terms on the HLPE-FSN from 2019 to 2023.

While I am not unbiased given my connections to the institution, I believe the HLPE-FSN/CFS is a rather remarkable Science–Policy–Society Interface given its transparent and inclusive nature. Clapp *et al.* (2023) have argued that legitimate food systems science–policy–society interfaces need to be independent,

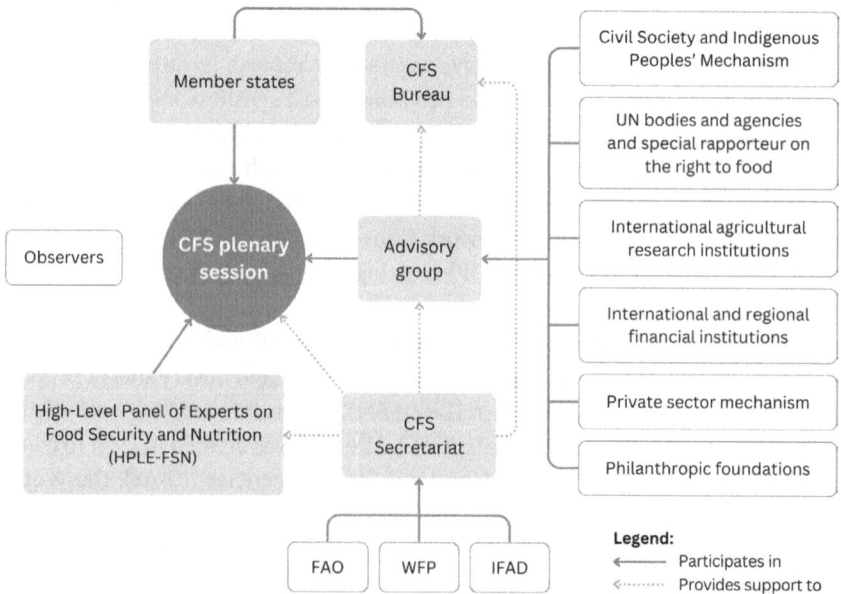

Figure 12.1 Organigram for the Committee on World Food Security (CFS) and its relationship to the High Level Panel of Experts for Food Security and Nutrition (HLPE-FSN)
Source: Figure design by Julia Castellano, Macalester College, source CSIPM (2024).

transparent, accessible, consultative and evidence based. The HLPE-FSN/CFS meets all of these criteria. It is from this SPSI that both agroecology and a six-dimensional approach to food security have been able to emerge and flourish in the UN system, a process that is beginning to bear fruit on the ground in Africa. Below I describe the general HLPE-FSN/CFS process, followed by a specific discussion about agroecology and the six-dimensional food security approach. I end by briefly commenting on the limitations of the HLPE-FSN/CFS.

Every four years the HLPE-FSN compiles a list of critical and emerging issues impacting global food security and nutrition that merit consideration. This list, as well as other concerns of CFS members, inform CFS political debates used to derive a four-year programme of work for the body. In most years, based on the aforementioned programme of work, the CFS commissions a report from the HLPE-FSN on a particular topic. The HLPE-FSN steering committee then organizes and oversees the writing of the report which is published and presented to the members of the CFS. The CFS then debates the recommendations of the report which are – in the normal order of work[2] – turned into voluntary guidelines. While these guidelines are not binding, they do have the potential to influence policy and programming within the UN system and at the level of member states. What is important to note is that at several points along the way in this process, there is space for public comment and input, including on (1) the original critical and emerging issues paper drafted by HLPE-FSN that influences the CFS agenda; (2) the CFS scoping brief that commissions a report; (3) and a draft of the report that the HLPE-FSN produces on a given topic. Then within the CFS (that debates and turns recommendations into guidelines), arguably the most direct representation of farmers and local food security advocates is the Civil Society and Indigenous Peoples Mechanism (CSIPM).

The CSIPM has regular representation in Rome (where the CFS meets) and reflects the interests and concerns of its regional constituents around the world, including farmer groups in Africa. More specifically, the CSIPM has a coordinating committee of 35 members representing 11 constituencies (e.g. herders, fisherfolk, food workers, etc.) and 17 subregions (of which there are five African subregional groups (North Africa, West Africa, East Africa, Central Africa and Southern Africa) (CSIPM 2023). So, for example, West Africa is represented on the CSIPM coordinating committee (as of late 2023) by Musa Sowe of the

---

2. I say normal order of work because in some instances the CFS has drafted voluntary guidelines that were not informed by a HLPE-FSN report on the topic. A good example of this are the CFS "Voluntary Guidelines on Gender Equality and Women's and Girls Empowerment in the Context of Food Security and Nutrition" (CFS 2023). Anecdotally, I note that most CFS members that I informally chatted with found it difficult to develop guidelines without a scientific report as a starting point.

Network of Farmer Organizations and Agricultural Producers of West Africa (ROPPA). The different regional groups hold consultations among themselves to determine their positions on different HLPE-FSN reports and their recommendations. Within the CSIPM, these positions then get discussed at an annual forum that happens before the annual CFS plenary in October every year where all of the groups on the CFS, country representatives, the private sector mechanism, CSIPM, and so on, hash out their differences and eventually develop voluntary guidelines. While there are a lot of levels to this process, it is more participatory than most international structures.

The emergence of agroecology

While the concept of agroecology can be traced in scientific literature back to the 1930s, and is increasingly intertwined with social movements from the 1990s (Wezel *et al.* 2009), it has largely been marginalized in mainstream agronomic research institutions and development bodies until recently. Loconto and Fouilleux (2019) argue that this began to change after the "Global Dialogue on Agroecology" was convened by the FAO in various cities around the world between 2014 and 2018 (namely Rome, Brasilia, Dakar, Bangkok, La Paz, Kunming and Budapest). Furthermore, they suggest that civil society organizations had a surprisingly high level of influence in these meetings in terms of how agroecology came to be understood and defined both as a science and a social movement.

As described in the CFS-HLPE-FSN process above, agroecology made its way into 2017 HLPE-FSN critical and emerging issues paper as a topic in need of further study (HLPE 2017). Perhaps more remarkably, given the different political interests represented at the CFS, agroecology was then selected by the CFS as a priority and it asked the HLPE-FSN to write a report on the topic that was released in 2019 (HLPE 2019). Based on this report, and amidst great controversy, the CFS released policy recommendations on "agroecological and other innovative approaches", a remarkable achievement given the longstanding peripheral position of this field in the scientific community (CFS 2021). The agroecology report and subsequent CFS recommendations have given agroecology an increased level of legitimacy within the UN system and more broadly. While it is still too early to chart the impact of these recommendations, it is already clear the increasing legitimacy of agroecological discourse at the international level (to which the CFS/HLPE-FSN report and recommendations contributed) has fostered a more permissive climate for agroecological innovations to flourish in diverse settings such as Senegal, Malawi, Brazil, India and France (Bezner Kerr 2020; Place *et al.* 2022). In some cases, there is now also funding for some of these initiatives, such as the European Commission funded FAIR project (Fostering Agroecological Intensification to improve farmers' Resilience) working in Mali, Burkina Faso and Senegal.

It is also worth mentioning that a 2020 HLPE-FSN report stressed the importance the six-dimensional food security framework discussed in Chapter 3 (which importantly added sustainability and agency as new dimensions of food security). While this framework never became an official CFS recommendation, the report had enough discursive power to begin influencing practices throughout the UN system, to the point where this framework is now used in the FAO's annual State of Food Security and Nutrition in the World (SOFI) reports. The report's recommendation has also gained widespread traction in the academic community (Clapp *et al.* 2022), to the point where the six-dimensional food security framework has now largely supplanted the four-dimensional conception.

Contemplating improvements to the CFS/HLPE-FSN

While I have described the CFS/HLPE-FSN as a promising science–policy–society interface because of its participatory processes and legitimacy, the body does have its limitations and could be improved. In my view, the HLPE-FSN would benefit from a greater level of autonomy. As it currently stands, the HLPE-FSN produces reports in response to requests from the CFS. The advantage of such an arrangement is that there is a built-in political audience for any report produced by the HLPE-FSN. This is also an advantage that HLPE-FSN has over other science-policy interfaces, such as the International Panel of Experts on Sustainable Food Systems (IPES-Food), which has discursive power, but no explicit connections to a policy making body. The downside of this arrangement is that the HLPE-FSN has limited autonomy to address new issues and crises that emerge rapidly (and outside of the four-year CFS workplan). The HLPE-FSN has been able to partially adapt in that it can now produce short issue papers at the request of the CFS. The HLPE-FSN successfully did this when it wrote three editions of an issue paper on the Covid-19-related food crisis that were well received and influential (HLPE 2021). This is, however, more complicated when politics within the CFS interfere. A good example of this is an issue paper that the HLPE-FSN wrote on food security implications of the war in Ukraine (HLPE 2022). It was challenging to get the CFS to ask the HLPE-FSN to write such a paper, which only happened when the CFS chair stuck out his neck and asked for it without the explicit approval of the CFS bureau. The CFS bureau, which included the representative from Russia, was not happy after the paper was released and blocked any subsequent updates as the situation in Ukraine, and the global food crisis, continued to evolve. Furthermore, there is also some disagreement on the CFS about how to respond more broadly to the 2020–23 global food crisis initially brought on by Covid-19 and then deepened by the war in Ukraine. Is the goal to return to a global food system as it was before 2019 or to build something different that is more decentralized, resilient and less productionist and trade-based? What this situation illustrates is that the politics of war (or some other nettlesome political disagreement) within the CFS, and the power of business interests working through some countries, can sometimes prevent the HLPE-FSN from doing its work. As such, the HLPE-FSN should have the autonomy to identify and write about emerging issues without an explicit request from the CFS.

A regressive counter-insurgency and attempted coup at the UN Food Systems Summit

While agroecology slowly gained currency within the UN system, this shift had not gone unnoticed. In fact, some members of the international agricultural establishment began to push back and attempted to marginalize the CFS/HLPE-FSN as the most inclusive and legitimate food security related science–policy–society interface in the UN system. This began as a call to create an "IPCC for food", a seemingly innocuous proposition that sounded great if you did not know that such an institution already existed in the form of the CFS/HLPE-FSN (Von Braun & Kalkuhl 2015). Then the UN Secretary General called for a UN Food Systems Summit to be organized in New York in September 2021. Unfortunately, "[t]he leadership selected by the Secretary General also clearly aligned the UNFSS with corporate and philanthropic interests that have long sought to promote market-oriented and technologically-driven approaches to food and agricultural systems" (Montenegro de Wit *et al.* 2021: 154). At the centre of this summit was a science committee run in a top-down fashion and chaired by a director who again called for the creation of an IPCC for food (Von Braun *et al.* 2023). To make matters worse, the many committees created to inform the summit's work were less than inclusive and susceptible to conventional agronomic thinking and policy prescriptions. Eventually civil society erupted in protest and boycotted the Summit as it became increasingly clear that the UNFSS had been hijacked by corporate interests (Canfield *et al.* 2021). Because of these protests, the UNFSS came and left with a whimper. Furthermore, the idea of creating a new IPCC for food never came to pass in the face of stiff opposition (Moseley 2021b).

Conclusion

This chapter has described and discussed the international and regional institutional architecture that has supported, sustained and proliferated conventional agronomic thinking on the African continent, even when repeated global food crises shook the global and regional food systems, raising important questions about the viability of such models. The chapter also presented a case study of a UN science–policy–society interface where, because of its participatory and inclusive nature, new thinking on agroecology was able to emerge and slowly begin to spread within the UN system as well as provide support for fledgling initiatives in certain African countries.

Furthermore, the emergence and spread of agroecology in the UN system, as well as the attempted coup at the UNFSS that sought to squash it, underscores the importance of agroecology as both a science and a social movement. Had agroecology not had the backing and support of civil society, it more than likely would neither have emerged in the UN system, nor survived the attempts to marginalize it. Neither agronomy nor agroecology is apolitical but intertwined with interests of those who benefit from it. Tropical agronomy has long benefitted from certain business and scientific interests who work relentlessly to promote and finance their world view in international and regional fora. In a similar way, agroecology is also a system of food production that benefits certain sets of actors, namely smallholder farmers, the poor and their allies. This group is now also learning that they have power when they form alliances and work together.

13

CONCLUSION: AIDING AND ABETTING RADICAL TRANSFORMATION

At some point in the early 2000s I made a visit to one of my farming friends in southern Mali after having been away for a few years. My friend, Mr Coulibaly, had been one of the more prosperous farmers in his community, growing and selling cotton, the major cash crop in the region. Yet, riding into his farm on the outskirts of the village that day, I noticed that there was no cotton to be seen. What could have happened? Why this change? I was simultaneously dumbfounded and pleased to learn that Coulibaly had decided to stop growing cotton on his own accord. This was a family decision that he and his brother had arrived at after much discussion. He told me that they had "had it" with cotton. The low prices and their concerns about the damage the crop might be doing to their soils had been weighing on them for some time. Instead, they were now focusing on food crops, and they had made maize their cash crop to make some money to cover basic expenses. Moreover, in addition to multiple, often intercropped, food crops (maize, sorghum, millet, cowpeas, peanuts, squash), a large vegetable garden and a tree orchard, they had begun producing compost with a vengeance to fertilize their fields, employing a system of several large composting pits and penned animals to capture their manure. While Coulibaly did not frame it as such, they had basically switched over to an agroecological approach.

I returned to see Coulibaly in 2009 and 2014 and noticed that cotton was slowly creeping back into their crop repertoire. In chatting with him, and walking around his farm together, I could tell that he was not super happy about this because he still had a lot of concerns about cotton, but the institutional support for the crop was just too strong to ignore, as well as the paucity of similar supports for food crops. It is worth reiterating that the Coulibalys had arrived at the decision to not grow cotton on their own. They were neither supported by a social movement, a network of farmers, nor did they receive any advice or encouragement from an extension or research institutions. In fact, they were heavily criticized by the local extension agency for their decision to stop growing cotton. I believe Coulibaly[1] was on the right track, but it is harder for individual farmers to strike out in new directions on their own, or they risk backsliding, when they do not have a community behind them or the support of institutions. The radical transformation of food systems may start at the grass roots as individuals become frustrated with the status quo, but they often require support if the change is to take root and spread.

This concluding chapter begins by reiterating the book's major argument. I then outline a theory of change or a process by which I believe agricultural development practice on the continent could shift away from the dominant paradigm. This theory of change engages with many of the historical conditions and conceptual insights initially presented in Chapters 1–3. I then go through each of the four country case studies presented in this book, synthesizing the big takeaways from the paired chapters for each country that analysed the problems, and then reasons for hope, in each case. I end with a brief conclusion which returns to the questions I asked at the end of Chapter 1.

From production agriculture to agroecology: a theory of change

As I noted in the introductory chapter, the book's major argument consists of three parts. These were as follows. First, development organizations and governments will only begin to seriously address hunger in Africa when they more fully question the assumption that increased crop production, using high external input agriculture, is the solution (an idea that is central to traditional agronomy). Second, agriculture development must be seen as more than the first step in an industrial development process but as a sustainable livelihood that has value in and of itself. And third, an agroecological approach, combined with good governance, will allow people to have greater control over their food systems, produce healthy food more sustainably and enhance access to food by the poorest of the poor.

Given the evidence and different situations this book has analysed in Mali, Burkina Faso, Botswana and South Africa, hopefully this argument resonates with the reader even more now than it did at the start of the book. However, a key question is how does one get there? While individual farmers making these decisions on their own is ultimately what must happen, it is often useful to have a support community (something the Coulibaly family was missing). Furthermore, ideally there would also be supportive government policies

---

1. Mr Coulibaly passed away in late 2023 at about the age of 85. He was a skilled farmer with a keen intellect and a kind heart. May his soul rest in peace.

or at least a government that was not actively trying to push farmers in the opposite direction. Here I present a theory of change, or the key ingredients needed for a radical transformation of African agriculture and food systems. As outlined in Figure 13.1 at the end of this section, five key ingredients (a crisis, a trenchant critique of the status quo, viable alternatives, social movements and engaged civil society, and institutional allies) are useful for bringing about a radical transformation of African agriculture and food systems and moving them towards more agroecological and six-dimensional food security futures. I discuss each of these ingredients in turn after briefly describing the status quo.

Status quo

The status quo agricultural development paradigm in the African context has been production agriculture since the colonial period. While the policies and practices in the colonial period were overtly colonial, and often enacted by force, what emerged in this period was a tropical agronomy that emphasized production-oriented agriculture for the benefit of the colonizing nations. While African countries experienced political decolonization from the 1960s, tropical agronomy was repackaged as development agronomy, and remains a force in African development circles today. Furthermore, there is a development agronomy world view that is embodied in a large number of African civil servants (certainly within African ministries of agriculture) that have been trained in this field.

Crisis, potentially weakened status quo

As discussed in Chapter 2 and outlined in Table 2.1, global and African food systems have experienced multiple food crises in the post-Second World War period, from concerns about population growth and food supplies in the 1960s and 1970s, to low commodity prices, high energy costs and the Third World Debt crisis of the late 1970s, to the global food crisis of 2007–08. Each time the status quo agricultural development machine has doubled down with a productionist response to these problems, from the first green revolution of the 1960s and 1970s, to structural adjustment and trade-based food security in the 1980s and 1990s, to neoproductionism and the New Green Revolution for Africa in the 2010s. The food crisis of 2020–23, brought on by Covid-19-related supply chain disruptions and the war in Ukraine has yet to be resolved, but it has created an opening for a different agricultural development paradigm.

In each case, these crises created an opening for change because crises often highlight the problems or weaknesses with the status quo. In the case of the first three crises, the international community was sufficiently alarmed that it rallied to reinforce the productionist agricultural model, often finding support among like-minded African civil servants who had been trained in agronomy. The opening created by the 2020–23 food crisis is still unresolved and it has led to many, many questions being raised about the status quo (Moseley & Battersby 2020; Moseley 2022a; Wudil *et al.* 2022).

Trenchant critique of the status quo

While crises may raise questions about the existing status quo, one also needs a coherent critique of the existing arrangements if policy-makers, farmers and broader publics are to consider a change. Intellectual colonization, political agronomy and political ecology highlight the role that development agronomy, international institutions, powerful corporate actors and certain states have played in building African food systems that are increasingly enmeshed in a global food system that is fragile and vulnerable to shocks (Clapp & Moseley 2020). A central focus of this critique is the productionist paradigm alleging that growing more food is the best way to address hunger and malnutrition (repeatedly asserted during the first wave of the Green Revolution and then again with the New Green Revolution for Africa), a claim laid bare as inadequate by a six-dimensional approach to food security (HLPE 2020). Furthermore, the energy intensive production methods introduced by this paradigm have created new vulnerabilities for African farmers (in terms of input supply shocks) and created a host of environmental problems.

Viable alternatives

A viable alternative is also necessary if African policy-makers, farmers and broader publics are to consider a change. At a broad level, a six-dimensional understanding of food security moves policy-makers away from a narrow focus on food production and availability, pushing them to start considering food access, utilization, stability, sustainability and agency. Agroecology also offers a coherent strategy for producing sufficient levels of healthy food in a sustainable manner. Lastly, agroecology also is an explicitly decolonial approach that bridges the two worlds of knowledge production: the experiential knowledge of farmers built up over generations; and the formal knowledge of the scientific community that is generated by study and experimentation.

Social movements and engaged civil society

If any new paradigm is to emerge and be sustained, it needs the active support of civil society and social movements. While some ideas arise from the grassroots, other paradigms may be more top down. Yet, if either are to be successful, they need a coalition of support that includes civil society. While emerging from the top, the productionist paradigm as a solution to hunger is engrained in the minds of broader publics in the West as it became a mantra in the post-Second World War era (e.g. American farmers are feeding the world). Agronomists, business interests and institutions were very skilled at cultivating public support by building and supporting a nearly hegemonic discourse supporting this way of thinking by idolizing heroic agronomy figures like Norman Borlaug and celebrating his legacy annually via World Food Day and other events (Quinn 2014; Stone 2018).

With the possible exception of South Africa, broader public support for the productionist solution to malnutrition has never been as deep in most African countries as it is in the West. Certainly among African farmers, there has always been a significant group that has approached cultivation in a way that is consistent with agroecology as a form of risk management and survival. Or, even if they were producing for the market, many African farmers have kept one leg in the subsistence world as a form of insurance. It is from this reservoir of subsistence which emerged food sovereignty, a social movement in support of agroecology and broader conceptions of food security (such as the six-dimensional approach). While this movement has yet to become fully widespread on the African continent, it has a critical role to play in helping to overturn the current paradigm in favour of agroecology.

Institutional allies

Even with a weakened status quo, a solid critique of the existing system, a viable alternative, and the support of social movements, it is extremely helpful to have institutional allies at the national, regional and/or international levels. International and regional partners may conduct research in support of agroecology, offer policy guidance, or supply funding to national governments for agroecology initiatives and education. Having allies within national governments is also key. As discussed in Chapters 3 and 5, when Thomas Sankara was president of Burkina Faso in the 1980s, his support for agroforestry and agroecology was key. Furthermore, and more recently, Senegalese Minister of Agriculture (now ambassador to Italy), Papa Abdoulaye Seck's epiphany about agroecology was an important step in his country's move away from production

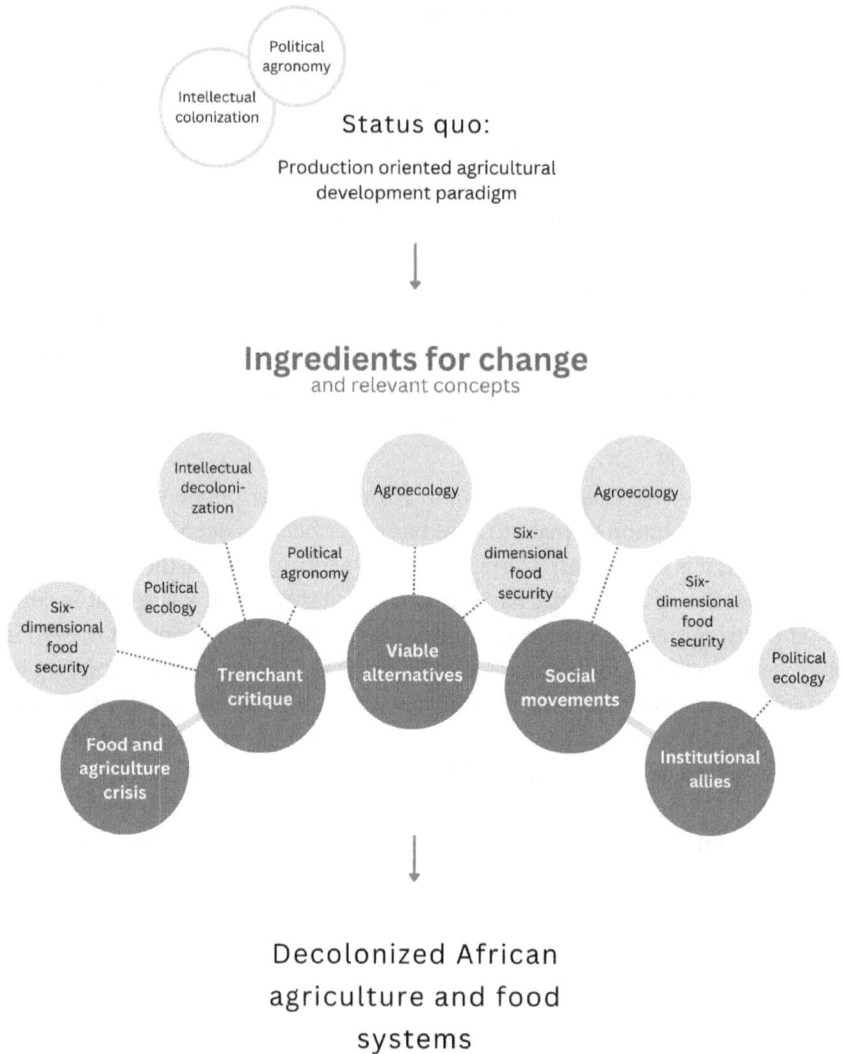

Figure 13.1 Decolonizing African agriculture: theory of change
Source: Figure conception by author, drawn by Julia Castellano, Macalester College.

agriculture as the solution to hunger (also discussed in Chapter 3). At the very least, one needs a national government that is not openly hostile to agroecology. This need for institutional allies is consistent with a political ecology approach that pushes one to think across multiple scales, from the local to the global. Unlike the productionist solution to hunger, which never had a firm connection to the grass roots in many African countries, and tended to be implemented in a more top down manner, agroecology is starting from the grassroots and needs to percolate into institutions at multiple scales if it is to be successful.

Country lessons

At the centre of this volume were eight chapters with my gleanings from four countries where I have spent most of my academic career working: Mali, Burkina Faso, Botswana and South Africa. The country studies in Part II of the book offered a critique of the existing situation, whereas the paired studies in Part III offered examples of alternative scenarios that hold more promise. In both Parts II and III, I drew on the concepts from part I (intellectual colonization/decolonization, political ecology, political agronomy, six-dimensional food security and agroecology) to offer a critique of the status quo and then to offer alternatives. Part IV of the book, looked beyond the country studies to examine the institutional architecture at the regional and international levels, some of which actively maintains the status quo and other cases where a new paradigm is emerging.

Mali

As elaborated in Chapter 4, we saw that cotton became increasingly important in Malian farming systems as a cash crop for export from the colonial period moving forward. Although once a crop largely grown by women for local weavers during the colonial period, the French largely destroyed the local weaving industry by flooding it with cheap textiles from abroad. While the French initially struggled to introduce new varieties of cotton as an export crop, this really started to take off in the late colonial period as the French moved away from the idea of irrigated cotton in the central part of the country and towards rainfed cotton in the south. This process was aided and abetted by taxation policies that made earning some cash necessary. The focus on cotton continued in the postcolonial period and especially following the World Bank's neoliberal economic reforms of the 1980s that emphasized exports. Subsequent research has demonstrated that cotton production has been deeply detrimental to soils and the environment more broadly. Furthermore, the wealthiest areas of the country that grow the largest amounts of cotton also suffer from the highest rates of child malnutrition. Mali's export led path to agricultural development has not led to improved food security, a direction supported by Malian government policy, French agronomic expertise and pressure from the World Bank and IMF. These trends are consistent with the insights of structural political ecology (which stresses core-periphery relationships) and political agronomy discussed in Chapter 3.

Yet, in the midst of these troublesome food security trends in Mali, there were also signs of hope (as discussed in Chapter 8). Mali's urban consumers continued to prefer local grains over imported ones, despite the proliferation of less nutritious, broken rice from Asia. Malian cotton farmers went on strike when producer prices were too low (as in 2000) or simply opted to grow less cotton when the conditions were not favourable (as in 2007–08 and 2020). Perhaps more interesting, Malian farmers chose to grow more food crops when they grew less cotton, a situation that saved the country from considerable social unrest in 2007–08 when the price of imported rice went up 100 per cent (as occurred in other countries in the region). This upsurge in food self-sufficiency was not only aided by labour organizing among farmers but a nascent food sovereignty movement that became increasingly well known globally after a landmark international meeting in the village of Nyéléni in southern Mali. This led to the Nyéléni Declaration, a statement of core principles of food sovereignty that align well with agroecology. The upsurge in sorghum production that accompanied declines in cotton is notable because sorghum requires fewer inputs, is more drought tolerant, is more amenable to intercropping and is more nutritious. Growing more food at home using agroecological methods, rather than being dependent on exports and a highly monetized food economy, may also make more sense in a country characterized by a weak state and increasing levels of political insecurity. Agency and social movements are a key element of agroecology, six-dimensional understandings of food security, and post structural political ecology.

Burkina Faso

As discussed in Chapter 5, while Burkina Faso has duelling agricultural traditions (export oriented agriculture versus agroecology), it has been an important site for the New Green Revolution for Africa (GR4A) in recent years. GR4A is a reboot of the original Green Revolution emphasizing improved seeds, pesticides and fertilizers, but with a new twist stressing public–private partnerships and value chains (a linear model for conceptualizing the different actors connecting African farmers to regional and global markets). In southwestern Burkina Faso, we explored the impacts of a GR4A rice project on the nutrition of participating women farmers as an example of big D development.[2] Even though this project

---

2. The concept of big D versus small d development is from Gillian Hart (2001) and Vicky Lawson (2007) and was discussed in Chapter 1. It basically refers to formal development efforts, or development programmes (big D development), versus background levels of economic change happening all the time.

was focused on a food crop (unlike cotton in Mali), it did not improve the food security and nutrition of women farmers, although it was good for the agronomists and seed companies involved (which was telling). While not directly connected to the rice project, the proliferating use of herbicides by women farmers in southwestern Burkina Faso is alarming (and a good example of little d development). I offered a multi-scaler and gendered political ecology explanation for this uptick in herbicide use related to: women's limited ability to commandeer labour within households; rural labour scarcity due to the departure of young people to work in artisanal gold mines; and increasingly affordable herbicides (including glyphosate), that are now produced generically in India and China. As such, affordable herbicides solve a labour problem for women, but the long-term health consequences for the community could be devastating.

Agroecology pushes one to think about the broader food environment, the area from which all food is gathered. In Chapter 9, I shared research on the value of foraging for dietary diversity in southwestern Burkina Faso, especially during the hungry season when other sources of food are scarce. This activity is largely conducted by women around farm fields and in semi-forested grasslands throughout the year. My students and I found that increased foraged food consumption was associated with higher measures of food security and dietary diversity after adjusting for wealth. While this type of activity is almost invisible to the formal agricultural development establishment, which is preoccupied with certain crops in farm fields, and often assumed to be a relic of the past, it is alive and well and is an incredibly important side activity that supports food security.

Botswana

As outlined in Chapter 6, semi-arid Botswana is a middle-income country with a well-managed diamond export industry that largely covers the expenses of the government and has allowed the country to build up substantial cash reserves. It also exports beef and has a successful ecotourism industry. In many ways, Botswana is an economic success story and is sometimes touted as an African development miracle. However, despite Botswana seemingly doing everything right (according to free trade advocate David Ricardo) by focusing on those sectors for which it has a comparative advantage and trading for food, the country experiences deep income inequality and alarmingly high levels of food insecurity for a middle-income country. Our research found that in the region near the capital in southeastern Botswana, 76 per cent of rural households, 67 per cent of peri-urban and 55 per cent of urban residents were facing moderate to severe food insecurity. The sources of this problem are structural in nature as the major pillars of the economy (diamonds, cattle and ecotourism) are driving inequality and poor households are vulnerable to high and fluctuating food prices. The situation was made worse by the government's decision to abandon food self-sufficiency as a policy goal in the 1980s. While this may have seemed rational from a neoliberal economic standpoint, it made less sense if one considers the food security and poverty alleviation benefits of small-scale agriculture.

In Chapter 10, Botswana's short-lived experiment with backyard gardening was examined. As a semi-arid country increasingly buffeted by climate change and dealing with persistent poverty and inequality, the country's decision to promote small-scale, irrigated horticulture made a lot of sense. This programme was also important because it was one of the few to support a largely female agricultural activity. Despite being condemned as a failure, we found that women's access to an affordable water supply was hugely important for the success of their gardens. In fact, those women who could tap into river water at relatively little cost could afford to be less commercially oriented and they had the lowest levels of food insecurity when they retained much of their vegetable production for home consumption (rather than being commercially oriented). In contrast, those women who had to pay for expensive tap water had to be commercially oriented to cover these costs and saw minimal food security improvements. Bigger picture, the insights of feminist political ecology were important for highlighting the relevance of socially constructed gender roles in initiatives to address food insecurity. The gardening programme actually showed a lot of promise and could be more effective if the government began to address the systematic male bias of its agricultural development activities. This bias has led to improvements in water access for the male dominated cattle industry and is a significant blind spot when it comes to addressing the water needs of women horticulturalists.

South Africa

As detailed in Chapter 7, in post-Apartheid South Africa one of the most significant initiatives of the new government was land reform. While the programme initiated under Nelson Mandela (known as Settlement/Land Acquisition Grants (SLAG)) had its problems, the programme was somewhat flexible about how land could be used by recipients and had a strong focus on poverty alleviation. Under Thabo Mbeki, the programme was transformed into Land Reform for Agricultural Development (LRAD) and became much more commercially oriented. This change reflected a strong bias in South Africa for commercial agriculture being the only type of credible agriculture. The programme failed for a variety of reasons, including that it was extremely challenging for new black farm owners to engage in large-scale commercial agriculture. This was not surprising given the incredibly competitive nature of the global agricultural products market, the extensive capital and contacts needed to thrive in this arena and an emerging set of environmental issues associated with large-scale commercial agriculture (such as pesticide resistance).

In Chapter 11, I argued that South Africa's commercially oriented land reform model is broken, neither being successful at redistributing land nor alleviating intertwined poverty and food insecurity. As such, South Africa ought to consider other land reform approaches involving low cost agroecological methods and smaller scale farms. In practical terms, this means allowing large commercial farms to be divided into smaller plots to be managed by small groups, individuals or households. While this approach was permitted under the original land redistribution programme, it has been virtually non-existent since 2000.

I explored the food security benefits of alternative land reform arrangements in South Africa's Western Cape Province in three ways. First, fair trade was examined as a mechanism for attaching additional value to products produced on farms that treated workers well and were owned or co-owned by historically marginalized groups. Second, I presented research on agricultural activities in the so called "coloured homelands" or former mission stations in the province, finding that small-scale agriculture and gardening provided significant income and food security benefits. Interestingly, provincial agricultural officials do not even really consider these types of activities to be real agriculture. Lastly, I shared a study of one of the few cases in the Western Cape where a large commercial farm was purchased with land reform funds and then split up into smaller plots. The land reform beneficiaries demonstrated considerable agency in getting this new approach off the ground and faced much resistance from the agricultural establishment along the way. The early results suggested that this approach was beginning to produce benefits and improved the food security of the households involved. Collectively, I refer to these alternatives as an agrarian justice approach.

Conclusion

At the end of the first chapter of this book, I mentioned a series of questions that I had been wrestling with over the course of my 35 plus years thinking about African agriculture and food security. Guided by some of the concepts and theories outlined in the first part of the book, and after examining problematic agricultural development efforts, as well as signs of hope, and then reflecting on the international and regional institutional architecture, these are some of the answers I have arrived at.

I first asked: why is this continent so rich in human ingenuity and natural resources not better fed? The answer here, and as discussed in Chapter 2 and throughout the case studies, has to do with the development of export-oriented agriculture in the colonial period, a process that was continued in the postcolonial era. With the support of agronomists, and encouraged by government policies, African farmers have produced more agricultural goods for export, become more vulnerable to climate fluctuations and damaged the environment in the process, and seen less and less income over time. While some farmers have resisted this process when it was acutely unfavourable, trade-based agriculture has not been good for African food security.

I then asked: what type of thinking has predominantly shaped the way policymakers and practitioners seek to develop agriculture and address food insecurity in the African context? The productionist approach to agricultural development and food security (which is central to development agronomy) is clearly at the centre of this problem. Associated with this is a tendency to think more about the development of agriculture as a step in the development process, rather than as a food security solution (even though the latter is given much lip service in the discourse). While this type of thinking is enshrined in agronomy, and dates to the colonial period, it continues to be supported and promulgated by the private sector, certain members of the scientific community, and institutions at different scales. In terms of alternatives (my next question), I spent a considerable amount of this book elaborating on a broader conceptions of food security and agroecology as an alternative paradigm.

Part II of the book examined: why have some approaches failed in certain places and why? In each case, Mali, Burkina Faso, Botswana and South Africa, the dominant production agriculture approach has led countries astray and failed to improve food security. While this was probably most clear in Mali and Burkina Faso, even in Botswana (a development success story), a decision to back away from crop agriculture had disastrous consequences for the poor and women. Furthermore, land redistribution efforts in South Africa, which had objectives to address historical injustices, were conceptually hijacked by commercially oriented production agriculture.

I then asked: what are signs of hope in those places that could be encouraged as the global community seeks to eliminate hunger over the next decade? This question was largely addressed in Part III of the book, where I explored signs of hope in Mali, Burkina Faso, Botswana and South Africa. In most cases these were accidental victories, or unseen successes, that point the way towards a more promising alternative. These included Mali's flirtation with food sovereignty, the unseen nutritional importance of rural food environments and foraging in Burkina Faso, the value of gardening and water access for women horticulturalists in Botswana, and the largely unrecognized potential of small-scale farming as model for a different land redistribution approach in South Africa.

My final question concerned: what types of changes at the international level are needed for positive change and progress to play out at the local level? This was largely explored in Chapter 12 (Part IV of the book), which examined international and regional agriculture and food security related institutions. While many of these institutions were put in place to support the first Green Revolution, and have persisted since that time, I explored a more participatory and inclusive structure, the Committee on World Food Security (CFS) and associated High Level Panel of Experts for Food Security and Nutrition (HLPE-FSN) where agroecology and the six-dimensional approach to food security have been able to emerge and spread throughout the United Nations system. The participatory nature of this structure helps one understand why this type of thinking emerged here and why civil society organizations could play a meaningful role in the process. These increasingly accepted ideas and concepts are also beginning to change the way the international community is thinking about African food security, a process that may eventually influence national governments and support farmers on the ground.

The reader may notice that the answers to these questions also lead us to the theory of change outlined at the start of this chapter. Interestingly, I have worked on different pieces of this theory of change at different points in my career, sometimes undertaking studies that offer a critique of the status quo (as I have done throughout my career and in this book); sometimes working to develop a viable alternative, either as an NGO worker supporting grain banks in central Mali, or a scholar writing about broader conceptions of food security and agroecology, and sometimes working as an ally within international institutions. But, as I also noted in the introduction, I am an outsider to the African communities I have worked with and learned from. In many cases, it would be inappropriate for me to become involved in certain steps in the change process. Nonetheless, I am also involved, both because agroecology spans the worlds of formal science (where I am located) and the experiential knowledge of African farmers. Furthermore, as political ecology demonstrates, African food insecurity is not disconnected from other actions in the world, including the decisions of my own government, the foods I consume on a daily basis, or the clothes I wear to stay warm. We are all in this together and we all have a stake in addressing African food insecurity and malnutrition. Getting there means dismantling the final frontier in colonialism, rethinking the core concepts, assumptions and theories that have led us down the wrong path for so long, and then acting on a new road map that has been co-created in a more decolonial fashion.

REFERENCES

Adam, J. (1948). "Les reliques boisées et les essences des savanes dans le zone préforestière en Guinee française". *Bulletin de la Société Botanique Française* 98: 22–6.

Africanews (2022). "Mali records highest cotton production". *Africanews*. 20 March. https://www.africanews.com/2022/03/20/mali-records-highest-cotton-production//.

AgriSA (2000). *"AgriSA's views on and involvement in land reform".* 25 April. Pretoria: AgriSA

Akram-Lodhi, A. (2021). "The ties that bind? Agroecology and the agrarian question in the twenty-first century". *Journal of Peasant Studies* 48(4): 687–714.

Alliance for a Green Revolution in Africa (AGRA) (n.d.). *AGRA Focus Countries.* https://agra.org/focus-countries/.

Alliance for Food Sovereignty in Africa (AFSA) (n.d.). "Who we are". https://afsafrica.org/.

Altieri, M. *et al.* (2015). "Agroecology and the design of climate change-resilient farming systems". *Agronomy for Sustainable Development* 35(3): 869–90.

ANC (2017). *54th Annual Conference. African National Congress (ANC). Report and Resolutions.* https://cisp.cachefly.net/assets/articles/attachments/73640_54th_national_conference_report.pdf.

Andersson, J. & J. Sumberg (2016). "Knowledge politics in development-oriented agronomy". Introductory paper prepared for the conference on Contested Agronomy, 23–25 February, Institute of Development Studies, Brighton, UK.

Annan, K. (2007). *"Remarks on the launch of the Green Revolution for Africa".* World Economic Forum, Cape Town, South Africa, 14 June.

Archer, S. (2000). "Technology and ecology in the Karoo: a century of windmills, wire and changing farming practice". *Journal of Southern African Studies* 26(4): 675–96.

Arku, G. (2009). "Rapidly growing African cities need to adopt smart growth policies to solve urban development concerns". *Urban Forum* 20(3): 253–70.

Assemblée Nationale du Burkina Faso (2016). *"Rapport General de La Commission d'Enquête Parlementaire sur La Gestion Des Titres Miniers et la Responsabilité Sociale des Entreprises Minieres".* Ouagadougou.

Badiane, M. *et al.* (2002). *Cotton Sector Strategies in West and Central Africa.* International Monetary Fund.

Baker, K., A. Clark & P. Imperato (2023). "Mali". *Encyclopedia Britannica.* https://www.britannica.com/place/Mali.

Bassett, T. (2001). *The Peasant Cotton Revolution in West Africa: Côte d'Ivoire, 1880–1995*. Cambridge: Cambridge University Press.

Bassett, T. & W. Munro (2022). "Lost in translation: pro-poor development in the Green Revolution for Africa". *African Studies Review* 65(1): 8–15.

Batisani, N. & B. Yarnal (2010). "Rainfall variability and trends in semi-arid Botswana: implications for climate change adaptation policy". *Applied Geography* 30: 483–9.

Battersby, J. & V. Watson (2018). *Urban Food Systems Governance and Poverty in African Cities*. Abingdon: Routledge.

Bebbington, A. & J. Carney (1990). "Geography in the international agricultural research centers: theoretical and practical concerns". *Annals of the American Association of Geographers* 80(1): 34–48.

Becker, L. (1994). "An early experiment in the reorganisation of agricultural production in the French Soudan (Mali), 1920–40". *Africa* 64(3): 373–90.

Becker, L. (2013). "Land sales and the transformation of social relations and landscape in peri-urban Mali". *Geoforum* 46: 113–23.

Becker, L. & N. Yoboué (2009). "Rice producer–processor networks in Côte d'Ivoire". *Geographical Review* 99(2): 164–85.

Bek, D., C. McEwan & K. Bek (2007). "Ethical trading and socioeconomic transformation: critical reflections on the South African wine industry". *Environment and Planning A* 39: 310–19.

Bélime, E. (1925). *La Production du Coton*. Paris: Comité Niger.

Bélime, E. (1941). *Les Travaux du Niger*. Paris: Martin-Mamy Crouan & Roques.

Berg, E. *et al.* (1994). *Poverty and Structural Adjustment in the 1980s: Trends in Welfare Indicators in Latin America and Africa* (USAID No. 2272–2019–4075). Washington, DC: US Agency for International Development.

Bertrand R., B. Keita & M. N'Diaye (1993). "La dégradation des sols des périmètres irrigués des grandes vallées Sud-saharienne. Cas de l'Office du Niger au Mali". *Cahiers Agricultures* 2(5): 318–29.

Besada, H. & B. O'Bright (2018). "Policy impacts on Africa's extractive sector: Botswana, diamond dependence, and diversification in the post-diamond period". *Revue Gouvernance* 15(2): 86–105.

Bezner Kerr, R. (2014). "Lost and found crops: agrobiodiversity, indigenous knowledge, and a feminist political ecology of sorghum and finger millet in northern Malawi". *Annals of the Association of American Geographers* 104(3): 577–93.

Bezner Kerr, R. (2020). "Agroecology as a means to transform the food system". *Landbauforschung* 70: 77–82.

Bezner Kerr, R. *et al.* (2019). "Repairing rifts or reproducing inequalities? Agroecology, food sovereignty, and gender justice in Malawi". *Journal of Peasant Studies* 46(7): 1499–518.

Blaikie, P. (1985). *The Political Economy of Soil Erosion in Developing Countries*. Harlow: Longman.

Boillat, S., R. Belmin & P. Bottazzi (2022). "The agroecological transition in Senegal: transnational links and uneven empowerment". *Agriculture and Human Values* 39(1): 281–300.

Bonneuil, C. & M. Kleiche (1993). *"Du jardin d'essais colonial à la station expérimentale 1880–1930: éléments pour une histoire du CIRAD"*. Montpellier: Centre de coopération internationale en recherche agronomique pour le développement (CIRAD).

Bret, P. (2002). "L'État, l'armée, la science. L'invention de la recherche publique en France (1763–1830)". *Annales historiques de la Révolution française* 328(1): 278.

Briney, A. (2019). "Overview of the Haber-Bosch process". https://www.thoughtco.com/overview-of-the-haber-bosch-process-1434563.

Brooks, A. & D. Simon (2012). "Unravelling the relationships between used-clothing imports and the decline of African clothing industries". *Development and Change* 43(6): 1265–90.

Brugger, F. & J. Zanetti 2020. "'In my village, everyone uses the tractor': gold mining, agriculture and social transformation in rural Burkina Faso". *The Extractive Industries and Society* 7(3): 940–53.

Bryant, C., B. Stephens & S. MacLiver (1978). "Rural to urban migration: some data from Botswana". *African Studies Review* 21(2): 85–99.

Bundy, C. (1979). *The Rise and Fall of the South African Peasantry*. Berkeley, CA: University of California Press.

Büscher, B. & R. Fletcher (2019). "Towards convivial conservation". *Conservation and Society* 17(3): 283–96.

Bush, R. (2010). "Food riots: poverty, power and protest". *Journal of Agrarian Change* 10(1): 119–29.

Canfield, M., M. Anderson & P. McMichael (2021). "UN Food Systems Summit 2021: dismantling democracy and resetting corporate control of food systems". *Frontiers in Sustainable Food Systems* 5: 103.

Canós-Donnay, S. (2019). "The empire of Mali". In *Oxford Research Encyclopedia of African History*. https://digital.csic.es/bitstream/10261/251370/1/Cano_s-Donnay 2018_Mali%20Empire.pdf.

Carney, J. (1993). "Converting the wetlands, engendering the environment: the intersection of gender with agrarian change in the Gambia". *Economic Geography* 69(4): 329–48.

Carney, J. (2002). *Black Rice: The African Origins of Rice Cultivation in the Americas*. Cambridge, MA: Harvard University Press.

Carroll, C., J. Vandermeer & P. Rossett (1990). *Agroecology*. New York, NY: McGraw-Hill.

Catling, D. (1996). "The 'rural areas' (landelike gebiede): their current status and development potential". *A socio-economic study prepared on behalf of Independent Development Trust*, Cape Town.

CEIC Data (2023a). *Botswana Foreign Exchange Reserves*. CEIC Data. https://www.ceicdata.com/en/indicator/botswana/foreign-exchange-reserves (accessed 18 August 2023).

CEIC Data (2023b). *United Kingdom Foreign Exchange Reserves*. CEIC Data. https://www.ceicdata.com/en/indicator/united-kingdom/foreign-exchange-reserves (accessed 18 August 2023).

Chappell, M. (2018). *Beginning to End Hunger: Food and the Environment in Belo Horizonte, Brazil, and Beyond*. Berkeley, CA: University of California Press.

Chevalier, A. (1909). *Rapport sur les nouvelles recherches sur les plantes à caouchouc de la Guinee Française*. IG276. Dakar: Archives du Senegal.

Chinsinga, B. (2012). *The Political Economy of Agricultural Policy Processes in Malawi: A Case Study of the Fertilizer Subsidy Programme*. Working paper #39. Future Agricultures.

Civil Society and Indigenous Peoples Mechanism (CSIPM) (2023). "What is the CSIPM?" https://www.csm4cfs.org/what-is-the-csm/#the-csm-forum.

Clapp, J. (2021). "Explaining growing glyphosate use: the political economy of herbicide-dependent agriculture". *Global Environmental Change* 67: 102239.

Clapp, J. & W. Moseley (2020). "This food crisis is different: COVID-19 and the fragility of the neoliberal food security order". *Journal of Peasant Studies* 47(7): 1393–417.

Clapp, J. *et al.* (2022). "The case for a six-dimensional food security framework". *Food Policy* 106: 102164.

Clapp, J. *et al.* (2023). "The I-TrACE principles for legitimate food systems science-policy-society interfaces". *Nature Food* 4: 128–29.

Clark, G. (1994). *Onions Are My Husband: Survival and Accumulation by West African Market Women*. Chicago, IL: University of Chicago Press.

Clarke, J. (2020). *The Wines of South Africa*. Oxford: Infinite Ideas.

Coates, J., A. Swindale & P. Bilinsky (2007). *Household Food Insecurity Access Scale (HFIAS) for Measurement of Food Access: Indicator Guide*. Washington, DC: US Agency for International Development.

Collier, P. (2007). *The Bottom Billion*. New York, NY: Oxford University Press.

Committee on Word Food Security (CFS) (2021). *Policy Recommendations for Agroecological and other Innovative Approaches for Sustainable Agriculture and Food Systems That Enhance Food Security and Nutrition*. https://www.fao.org/fileadmin/templates/cfs/Docs2021/Documents/Policy_Recommendations_Agroecology_other_Innovations/2021_Agroecological_and_other_innovations_EN.pdf.

Committee on Word Food Security (CFS) (2023). *Voluntary Guidelines on Gender Equality and Women's and Girls Empowerment in the Context of Food Security and Nutrition*. https://www.fao.org/fileadmin/templates/cfs/Docs2223/Gender/Guidelines_Final_Agreed_Version_June_2023_CLEAN/GEWGE_Guidelines_Final_Agreed_Version_June_2023_CLEAN.pdf.

Conrad, D. (2009). *Great Empires of the Past: Empires of Medieval West Africa: Ghana, Mali, and Songhay*. New York, NY: Chelsea House.

Consultative Group on International Agricultural Research (CGIAR) (2022). *CGIAR's Agroecology Initiative: Transforming Food, Land, and Water Systems Across the Global South*. https://www.cgiar.org/news-events/news/cgiars-agroecology-initiative-transforming-food-land-and-water-systems-across-the-global-south/.

Consultative Group on International Agricultural Research (CGIAR) (2023a). "Research centers". https://www.cgiar.org/research/research-centers/.

Consultative Group on International Agricultural Research (CGIAR) (2023b). *CGIAR Financial Report Dashboards*. https://www.cgiar.org/food-security-impact/finance-reports/dashboard/center-analysis/.

Cousins, B., (2006). "Land and agriculture policy: back to the roots". *Mail & Guardian*, 15 August. https://mg.co.za/article/2006-08-15-land-and-agriculture-policy-back-to-the-roots/.

Cronon, W. (1996). "The trouble with nature or, getting back to the wrong wilderness". *Environmental History* 1(1): 7–28.

CSIPM (2024). "What is the CSIPM?" Civil Society and Indigenous Peoples' Mechanism of the UN Committee on World Food Security. https://www.csm4cfs.org/what-is-the-csm/ (accessed 16 May 2024).

Dabiré, K. *et al.* (2012). "Trends in insecticide resistance in natural populations of malaria vectors in Burkina Faso, West Africa: 10 years' surveys". *Insecticides-Pest Engineering* 22: 479–502.

Daily Investor (2023). "Very difficult time ahead for South Africa". *Daily Investor*, 7 August. https://dailyinvestor.com/south-africa/26560/very-difficult-time-ahead-for-south-africa/.

Dalrymple, D. (1979). "The adoption of high-yielding grain varieties in developing nations". *Agricultural History* 53(4): 704–26.

Darkoh, M. & J. Mbaiwa (2002). "Globalisation and the livestock industry in Botswana". *Singapore Journal of Tropical Geography* 23(2): 149–66.

Davies, S. (2016). *Adaptable Livelihoods: Coping with Food Insecurity in the Malian Sahel*. Berlin: Springer.

Davis, M. (2002). *Late Victorian Holocausts: El Niño Famines and the Making of the Third World*. London: Verso.

Dembele, D. (2013). "Thomas Sankara: an endogenous approach to development". *Pambazuka News*, 23 October. https://www.pambazuka.org/pan-africanism/thomas-sankara-endogenous-approach-development.

De Schutter, O. (2014). "The reform of the Committee on World Food Security: the quest for coherence in global governance". In N. Lambek *et al.* (eds), *Rethinking Food Systems: Structural Challenges, New Strategies and the Law*, 219–38. Berlin: Springer.

de Wet, J. & J. Huckabay (1967). "The origin of sorghum bicolor. II. Distribution and domestication". *Evolution* 21(4): 787–802.

Delgado, C. & C. Miller (1985). "Changing food patterns in West Africa: implications for policy research". *Food Policy* 10(1): 55–62.

Diamond, J. (1987). "The worst mistake in the history of the human race". *Discover Magazine*, May.

Diarra, A. (2015). "Revue des politiques sur les pesticides et les produits vétérinaires dans l'espace CEDEAO". *Laboratoire d'innovation FSP – Document de Travail N° West Africa-JSR-2015–2*. East Lansing, MI: Michigan State University.

Dowd, B. (2008). "A march to a better future for Africa's poor? The introduction of Bt cotton in Burkina Faso". *African Geographical Review* 27(1): 17–21.

Du Toit, A. (1993). "The micro-politics of paternalism: the discourse of management and resistance on South African fruit and wine farms". *Journal of Southern African Studies* 19(2): 314–36.

Du Toit, A. (1994). "Farm workers and the 'agrarian question'". *Review of African Political Economy* 21(61): 375–88.

Du Toit, P. (2004). *The Great South African Land Scandal*. Centurion, SA: Legacy Publications.

Du Toit, A & J. Ewert (2002). "Myths of globalisation: private regulation and farm worker livelihoods on Western Cape farms". *Transformation 50*: 77–104.

Duke, S. & S. Powles (2008). "Glyphosate: a once-in-a-century herbicide: mini-review". *Pest Management Science* 64(4): 319–25.

Dury, S. & I. Bocoum (2012). "Le 'paradoxe' de Sikasso (Mali): pourquoi 'produire plus' ne suffit-il pas pour bien nourrir les enfants des familles d'agriculteurs?" *Cahiers Agricoles* 21(5): 324–36.

Duvall, C. (2007). "Human settlement and baobab distribution in south-western Mali". *Journal of Biogeography* 34(11): 1947–61.

Echenberg, M. *et al.* (2023). "Burkina Faso". *Encyclopedia Britannica*. https://www. britannica.com/place/Burkina-Faso.

Economist Impact (2022). *The Global Food Security Index 2022*. https://impact.economist. com/sustainability/project/food-security-index/reports/Economist_Impact_ GFSI_2022_Global_Report_Sep_2022.pdf.

Ehrlich, P. (1968). *The Population Bomb*. San Francisco, CA: The Sierra Club.

Eicher, C. (1995). "Zimbabwe's maize-based green revolution: preconditions for replication". *World Development* 23(5): 805–18.

Eklin, K. *et al.* (2014). *The Committee on World Food Security Reform: Impacts on Global Governance of Food Security*. IDDRI working paper. https://hal.science/hal-01620862/ document.

Ellis, W. (1987). "Africa's stricken Sahel". *National Geographic Magazine* 172(2): 140–59.

Encyclopaedia Britannica (2023). "Freedom charter". https://www.britannica.com/topic/ Freedom-Charter.

Engels, B. (2015). "Contentious politics of scale: the global food price crisis and local protest in Burkina Faso". *Social Movement Studies* 14(2): 180–94.

Englebert, P. (1996). *Burkina Faso: Unsteady Statehood in West Africa*. New York, NY: Routledge.

ESRI Africa (2018). "GeoPortal. Africa countries". Esri Africa. https://www.africa geoportal.com/datasets/64aff05d66ff443caf9711fd988e21dd_0/explore?location=-3. 232759%2C19.067600%2C4.00 (accessed 13 May 2024).

Fairhead, J. & M. Leach (1995). "False forest history, complicit social analysis: rethinking some West African environmental narratives". *World Development* 23(6): 1023–35.

Fairhead, J. & M. Leach (1996). *Misreading the African Landscape: Society and Ecology in a Forest-Savanna Mosaic*. New York, NY: Cambridge University Press.

Fairtrade Labelling Organizations International (FLO) (2008a). "List of certified operators". http://www.flo-cert.net/flo-cert/operators.php?id=10.

Fairtrade Labelling Organizations International (FLO) (2008b). *Guidance Document for Fair-Trade Labeling: Standards Guidance for South Africa*. http://www.fairtrade.net/ fileadmin/user_upload/content/Guidance_South-Africa-30July04.pdf.

Food and Agriculture Organization (FAO) (2006). *Food Security*. Policy Brief, Issue 2. http://www.fao.org/forestry/13128-0e6f36f27e0091055bec28ebe830f46b3.pdf

Food and Agriculture Organization (FAO) (2012). *FAO Food Price Index.* Rome: Food and Agriculture Organization of the United Nations. https://www.fao.org/world foodsituation/foodpricesindex/esn.

Food and Agriculture Organization (FAO) (2018). *Botswana: National Gender Profile of Agriculture and Rural Livelihoods.* Gaborone: FAO. https://www.fao.org/3/i8704en/ I8704EN.pdf.

Food and Agriculture Organization (FAO) (2021). *The State of Food Security and Nutrition in the World 2021.* Rome: UN Food and Agriculture Organization.

Food and Agriculture Organization (FAO) (2023). *Cereal Supply and Demand Balances for Sub-Saharan African Countries: Situation as of June 2023.* Rome: Food and Agriculture Organization of the United Nations. https://doi.org/10.4060/cc7114en.

FAO & FHI 360 (2016). *Minimum Dietary Diversity for Women: A Guide for Measurement.* Rome: UN Food and Agriculture Organization.

Fehr, R. (2016). *Exploring the Role of Horticulture in Alleviating Food Insecurity Among Women in Botswana.* Geography Honors Thesis, Macalester College. https:// digitalcommons.macalester.edu/geography_honors/48.

Fehr, R. & W. Moseley (2019). "Gardening matters: a political ecology of female horti-culturists, commercialization, water access and food security in Botswana". *African Geographical Review* 38(1): 67–80.

Ferguson, J. (1994). *The Anti-Politics Machine: "Development", Depoliticization, and Bureaucratic Power in Lesotho.* Minneapolis, MN: University of Minnesota Press.

Fiamingo, C. (2021). "'Expropriation without compensation': una lotta a lance spuntate". *Afriche e Orienti,* 24(2): 1–26.

Filipovich, J. (2001). "Destined to fail: forced settlement at the Office du Niger, 1926–45". *Journal of African History* 42(2): 239–60.

Fok, M. (2000). *"Histoire du développement de la filière cotonnière au Mali: rôle et place des innovations institutionnelles".* Montpellier: CIRAD.

Forbes, R. (1933). "The black man's industries". *Geographical Review* 23(2): 230–47.

Forbes, R. (1943). "The trans-Saharan conquest". *Geographical Review* 33(2): 197–213.

Fouberg, E. & W. Moseley (2018). *Understanding World Regional Geography.* Second edn. Hoboken, NJ: Wiley/Blackwell.

Foucault, M. (1971). *The Archaeology of Knowledge.* New York, NY: Pantheon.

Foucault, M. (1980). *Power/Knowledge: Selected Interviews and Other Writings, 1972–77.* New York, NY: Pantheon.

Franke, R. & B. Chasin (1980). *Seeds of Famine: Ecological Destruction and the Development Dilemma in the West African Sahel.* New Jersey: Allanheld, Osmun.

Freire, P. (1982). *Pedagogy of the Oppressed.* New York, NY: Continuum.

Gazeley, J. (2022). "The strong 'weak state': French statebuilding and military rule in Mali". *Journal of Intervention and Statebuilding* 16(3): 269–86.

Gengenbach, H. *et al.* (2018). "Limits of the New Green Revolution for Africa: reconcep-tualising gendered agricultural value chains". *Geographical Journal* 184(2): 208–14.

GIEWS (2022). *GIEWS Country Brief – Botswana. Global Information and Early Warning System (GIEWS).* Rome: UN Food and Agriculture Organization. https://www.fao. org/giews/countrybrief/country/BWA/pdf/BWA.pdf.

Gilmore, R. (2020). *Geographies of Racial Capitalism with Ruth Wilson Gilmore*. K. Card (director) [Video/DVD]. Antipode Foundation. https://www.youtube.com/watch?v=2CS627aKrJI.

Global Nutrition Report (2023). *Botswana:Country Nutrition Profile*. https://globalnutritionreport.org/resources/nutrition-profiles/africa/southern-africa/botswana/.

Good, K. (1993). "At the ends of the ladder: radical inequalities in Botswana". *Journal of Modern African Studies* 31(2): 203–30.

Government of Botswana (GoB). (2010). *Feasibility Design Study on the Utilization of the Water Resources of the Chobe/Zambezi River*. Ministry of Minerals, Energy and Water Resources, Department of Water Affairs. https://www.water.gov.bw/images/Draft%20Prefeasibility_Feasibility%20Report.pdf.

Grabowski, P. & T. Jayne (2016). "Analyzing trends in herbicide use in Sub-Saharan Africa". *Michigan State University International Development Working Paper No. 142* (1096-2016-88368).

Grain (2007). "Nyéléni – for food sovereignty". https://grain.org/en/article/601-nyeleni-for-food-sovereignty.

Gramsci, A. (1971). *Selections from the Prison Notebooks*. Edited and translated by Q. Hoare & G. Nowell Smith. New York, NY: International Publishers.

Graves, N. (1965). "The 'Grandes Ecoles' in France". *The Vocational Aspect of Secondary and Further Education* 17(36): 40–49.

Gray, L. & W. Moseley (2005). "A geographical perspective on poverty–environment interactions". *Geographical Journal* 171(1): 9–23.

Green, T. & J. Thornton (2022). "Did European trade with Africans (including the slave trade) damage or ruin economies on the continent?" In W. Moseley & K. Otiso (eds), *Debating African Issues: Conversations Under the Palaver Tree*, 12–22. Agingdon: Routledge.

Green, T. (2019). *A Fistful of Shells: West Africa from the Rise of the Slave Trade to the Age of Revolution*. London: Penguin.

Grinde, M. (2008). *Small-Hold Alternatives for Land Redistribution Projects in the Western Cape*, South Africa: A Case Study. Geography Honors Project, Macalester College. https://digitalcommons.macalester.edu/geography_honors/14.

Gulati, A. & S. Fan (2007). *The Dragon and the Elephant: Agricultural and Rural Reforms in China and India*. Baltimore, MD: Johns Hopkins University Press.

Gwebu, T. (2003). "Environmental problems among low income urban residents: an empirical analysis of old Naledi-Gaborone, Botswana". *Habitat International* 27(3): 407–27.

Haggblade, S. *et al.* (2017a). "The herbicide revolution in developing countries: patterns, causes, and implications". *European Journal of Development Research* 29: 533–59.

Haggblade, S. *et al.* (2017b). "Causes and consequences of increasing herbicide use in Mali". *European Journal of Development Research* 29: 648–74.

Hall, R. (2004). "A political economy of land reform in South Africa". *Review of African Political Economy* 31(100): 213–27.

Hall, R. (2010). "Two cycles of land policy in South Africa: tracing the contours". In T Vircoulon, W. Anseeuw & C. Alden (eds), *The Struggle over Land in Africa Conflicts, Politics and Change*, 175–92. Pretoria, SA: HSRC Press.

Hall, R. (2012). "The next Great Trek? South African commercial farmers move north". *Journal of Peasant Studies* 39(3/4): 823–43.

Hall, R. & T. Kepe (2017). "Elite capture and state neglect: new evidence on South Africa's land reform". *Review of African Political Economy* 44(151): 122–30.

Hammer, J. (2016). "The librarian who saved Timbuktu's cultural treasures from Al Qaeda". *Wall Street Journal*, 15 April. https://www.wsj.com/articles/the-librarian-who-saved-timbuktus-cultural-treasures-from-al-qaeda-1460729998.

Hancock, G. (1989). *Lords of Poverty: The Power, Prestige, and Corruption of the International Aid Business*. New York, NY: Atlantic Monthly Press.

Hardin, G. (1968). "The tragedy of the commons". *Science* 162(3859): 1243–48.

Harrison, P. (1987). *The Greening of Africa*. New York, NY: Penguin.

Hart, G. (2001). "Development critiques in the 1990s: culs de sac and promising paths". *Progress in Human Geography* 25(4): 649–58.

Hayes, T. *et al.* (2002). "Hermaphroditic, demasculinized frogs after exposure to the herbicide atrazine at low ecologically relevant doses". *Proceedings of the National Academy of Sciences* 99(8): 5476–80.

Heap, I. (2014). "Global perspective of herbicide-resistant weeds". *Pest Management Science* 70(9): 1306–15.

Henderson, W. (1933). "The cotton famine on the continent, 1861–65". *Economic History Review* 4(2): 195–207.

Herbart, P. (1939). *Le Chancre du Niger*. Paris: Gallimard.

Hillbom, E. (2008). "Diamonds or development? A structural assessment of Botswana's forty years of success". *Journal of Modern African Studies* 46: 191–214.

Hillbom, E. (2011). "Botswana: a development-oriented gate-keeping state". *African Affairs* 111(442): 67–89.

HLPE (2017). *2nd Note on Critical and Emerging Issues for Food Security and Nutrition*. A note by the High Level Panel of Experts on Food Security and Nutrition of the Committee on World Food Security, Rome. https://www.fao.org/fileadmin/user_upload/hlpe/hlpe_documents/Critical-Emerging-Issues-2016/HLPE_Note-to-CFS_Critical-and-Emerging-Issues-2nd-Edition__27-April-2017_.pdf.

HLPE (2019). *Agroecological and Other Innovative Approaches for Sustainable Agriculture and Food Systems that Enhance Food Security and Nutrition*. A report by the High Level Panel of Experts on Food Security and Nutrition, UN Committee on World Food Security, Rome. https://www.fao.org/3/ca5602en/ca5602en.pdf.

HLPE (2020). *Food Security and Nutrition: Building A Global Narrative Towards 2030*. Report #15. High Level Panel of Experts for Food Security and Nutrition (HLPE-FSN), UN Committee on World Food Security (CFS). http://www.fao.org/3/ca9731en/ca9731en.pdf.

HLPE (2021). *Impacts of COVID-19 on Food Security and Nutrition: Developing Effective Policy Responses to Address the Hunger and Malnutrition Pandemic*. An issue paper by the High Level Panel of Experts on Food Security and Nutrition of the Committee on World Food Security, Rome. https://www.fao.org/3/ng808en/ng808en.pdf.

HLPE (2022). *The Impacts on Global Food Security and Nutrition of the Military Conflict in Ukraine*. An issue paper by the High Level Panel of Experts on Food Security and

Nutrition of the Committee on World Food Security, Rome. https://www.fao.org/fileadmin/templates/cfs/HLPE/reports/issues_paper/Impacts_conflict_Ukraine_FSN_HLPE_Issues_Paper.pdf.

Hope Sr, K. (1995). "Managing the public sector in Botswana: some emerging constraints and the administrative reform responses". *International Journal of Public Sector Management* 8(6): 51–62.

Hore, R. (2022). "P1bn monthly food import bill irks". August 15. *Botswana Business Weekly*, 15 August. https://businessweekly.co.bw/columns/guest-contributor/p1bn-monthly-food-import-bill-irks.

Hovorka, A. (2006). "The No. 1 Ladies' Poultry Farm: a feminist political ecology of urban agriculture in Botswana". *Gender, Place & Culture* 13(3): 207–25.

Hovorka, A. (2012). "Women/chickens vs men/cattle: insights on gender–species intersectionality". *Geoforum* 43(4): 875–84.

Humanitarian Data Exchange (2018). "Sahel: administrative boundaries and settlements". United Nations Office for the Coordination of Humanitarian Affairs. https://data.humdata.org/dataset/sahel-administrative-boundaries (accessed 15 May 2023).

Humanitarian Data Exchange (2020). "South Africa: subnational administrative boundaries". United Nations Office for the Coordination of Humanitarian Affairs. https://data.humdata.org/dataset/cod-ab-zaf (accessed 16 May 2023).

Humanitarian Data Exchange (2022). "Burkina Faso: settlements". United Nations Office for the Coordination of Humanitarian Affairs. https://data.humdata.org/dataset/settlements (accessed 15 May 2023).

Humanitarian Data Exchange (2023). "Burkina Faso: subnational administrative boundaries". United Nations Office for the Coordination of Humanitarian Affairs. https://data.humdata.org/dataset/cod-ab-bfa? (accessed 15 May 2023).

Hunter-Adams, J., J. Battersby & T. Oni (2019). "Food insecurity in relation to obesity in peri-urban Cape Town, South Africa: implications for diet-related non-communicable disease". *Appetite* 137: 244–9.

Huss-Ashmore, R. & S. Johnston (1997). "Wild plants as famine foods: food choice under conditions of scarcity". In H. Macbeth (ed.), *Food Preferences and Taste: Continuity and Change*, 83–100. Oxford: Berghahn.

Iacovides, I. (2017). "Artificial groundwater recharge practice in Cyprus". In I. Simmers (ed.), *Recharge of Phreatic Aquifers in (Semi-) Arid Areas*, 201–14. Abingdon: Routledge.

Intergovernmental Panel on Climate Change (IPCC) (2023). *AR6 Synthesis Report: Climate Change 2023*. https://www.ipcc.ch/report/sixth-assessment-report-cycle/.

Jacobs, R. (2013). "The radicalization of the struggles of the food sovereignty movement in Africa". In *La Via Campesina's Open Book: Celebrating 20 Years of Struggle and Hope*.

Jakobsen, J. & O. Westengen (2022). "The imperial maize assemblage: maize dialectics in Malawi and India". *Journal of Peasant Studies* 49(3): 536–60.

Jarosz, L. (2003). "A human geographer's response to guns, germs, and steel: the case of agrarian development and change in Madagascar". *Antipode* 35(4): 823–8.

Kadibadiba, T., L. Roberts & R. Duncan (2018). "Living in a city without water: a social practice theory analysis of resource disruption in Gaborone, Botswana". *Global Environmental Change* 53: 273–85.

Kansanga, M. *et al.* (2019). "Traditional agriculture in transition: examining the impacts of agricultural modernization on smallholder farming in Ghana under the new Green Revolution". *International Journal of Sustainable Development & World Ecology* 26(1): 11–24.

Kansanga, M. *et al.* (2022). "Time matters: a survival analysis of timing to seasonal food insecurity in semi-arid Ghana". *Regional Environmental Change* 22(2): 41.

Keiner, M. *et al.* (eds) (2013). *From Understanding to Action: Sustainable Urban Development in Medium-Sized Cities in Africa and Latin America* (Vol. 5). New York, NY: Springer.

Kennedy, G., T. Ballard & M.-C. Dop (2010). *Guidelines for Measuring Household and Individual Dietary Diversity.* Rome: FAO.

Kevane, M. (2015). "Gold mining and economic and social change in West Africa". In C. Monga & J. Lin (eds), *Oxford Handbook of Africa and Economics: Volume 2: Policies and Practices*, 240–53. Oxford: Oxford University Press.

Klein, M. (1998). "Review of *Two Worlds of Cotton: Colonialism and the Regional Economy of the French Soudan, 1800–1946*, by Richard L. Roberts". *Canadian Journal of History* 33(3): 488–90.

Konadu-Agyemang, K. (2000). "The best of times and the worst of times: structural adjustment programs and uneven development in Africa: the case of Ghana". *The Professional Geographer* 52(3): 469–83.

Kriger, C. (2005). "Mapping the history of cotton textile production in precolonial West Africa". *African Economic History* 33: 87–116.

Kruger, S. & A. Du Toit (2007). "Reconstructing fairness: fair trade conventions and worker empowerment in South African horticulture". In L. Raynolds, D. Murray & J. Wilkinson (eds), *Fair Trade: The Challenges of Transforming Globalization*, 216–36. London: Routledge.

Laband, J. (2020). *The Land Wars: The Dispossession of the Khoisan and AmaXhosa in the Cape Colony.* Cape Town: Penguin Random House South Africa.

Lacy, S. (2008). "Cotton casualties and cooperatives: reinventing farmer collectives at the expense of rural Malian communities". In W. Moseley & L. Gray (eds), *Hanging by a Thread: Cotton, Globalization, and Poverty in Africa*, 207–33. Athens, OH: Ohio University Press.

Lado, C. (2001). "Environmental and socio-economic factors behind food security policy strategies in Botswana". *Development Southern Africa* 18(2): 141–68.

LaFargeas, P. (2019). "Aux sources de l'agronomie tropicale". Le Blog Gallica. Bibliotheque Nationale de France. https://gallica.bnf.fr/blog/07062019/aux-sources-de-lagronomie-tropicale.

Laris, P. & J. Foltz (2014). "Cotton as catalyst? The role of shifting fertilizer in Mali's silent maize revolution". *Human Ecology* 42: 857–72.

Laris, P., J. Foltz & B. Voorhees (2015). "Taking from cotton to grow maize: the shifting practices of small-holder farmers in the cotton belt of Mali". *Agricultural Systems* 133: 1–13.

Laske, E. (2021). Mesure du contenu en emploi de l'agriculture en Afrique subsaharienne: comparaison des modèles agroécologique et conventionnel (Doctoral dissertation, Université de Montpellier).

Lawson, V. (2007). *Making Development Geography.* London: Hodder Arnold.

Leach, M. & R. Mearns (eds) (1996). *The Lie of the Land: Challenging Received Wisdom on the African Environment.* London: James Currey.

Lecompte, B. & A. Krishna (1997). "Six-S: building upon traditional social organizations in Francophone West Africa". In A. Krishna, N. Uphoff & M. Esman (eds), *Reasons for Hope: Instructive Experiences in Rural Development,* 75–90. Boulder, CO: Lynne Reinner.

Ledermann, S. & W. Moseley (2007). "The World Trade Organization's Doha round and cotton: continued peripheral status or a 'historical breakthrough' for African farmers?" *African Geographical Review* 26(1): 37–58.

Leshoele, M. (2019). *Pan-Africanism and African Renaissance in Contemporary Africa: Lessons from Burkina Faso's Thomas Sankara.* Doctoral dissertation, University of South Africa. https://uir.unisa.ac.za/bitstream/handle/10500/26595/thesis_leshoele_m.pdf.

Lester, A. (2000). "Historical geography". In R. Fox & K. Rowntree (eds), *The Geography of South Africa in a Changing World,* 60–85. Oxford: Oxford University Press.

Lewis, J. (1979). *Descendants and Crops: Two Poles of Production in a Malian Peasant Village.* PhD dissertation, Yale University.

Lewis, J. (2022). *Slaves for Peanuts: A Story of Conquest, Liberation, and a Crop that Changed History.* New York, NY: The New Press.

Lightfoot, D. (2000). "The origin and diffusion of qanats in Arabia: new evidence from the northern and southern peninsula". *Geographical Journal* 166(3): 215–26.

Loconto, A. & E. Fouilleux (2019). "Defining agroecology: exploring the circulation of knowledge in FAO's Global Dialogue". *International Journal of Sociology of Agriculture and Food* 25(2): 116–37.

Lohnes, K. (2018). "Battle of Blood River". *Encyclopedia Britannica.* https://www.britannica.com/event/Battle-of-Blood-River.

Lovejoy, P. (1989). "The impact of the Atlantic slave trade on Africa: a review of the literature". *Journal of African History* 30(3): 365–94.

Loxley, J. (1990). "Structural adjustment in Africa: reflections on Ghana and Zambia". *Review of African Political Economy* 17(47): 8–27.

Luna, J. (2020). "'Pesticides are our children now': cultural change and the technological treadmill in the Burkina Faso cotton sector". *Agriculture and Human Values* 37: 449–62.

Mackett, O. (2021). "Female farm holding in Botswana's agriculture industry". *Agrekon* 60(3): 317–34.

Maharaj, B. & M. Ramutsindela (2021). "Social change and the (re)radicalization of geography in South Africa". In L. Berg *et al.* (eds), *Placing Critical Geography: Historical Geographies of Critical Geography,* 28–43. Abingdon: Routledge.

Mandela, N. (1995). *A Long Walk to Freedom: The Autobiography of Nelson Mandela.* New York, NY: Time Warner.

Manning, P. (1981). "The enslavement of Africans: a demographic model". *Canadian Journal of African Studies/La Revue Canadienne Des Études Africaines* 15(3): 499–526.

Manning, P. (2004). *Slavery, Colonialism and Economic Growth in Dahomey, 1640–1960.* New York, NY: Cambridge University Press.

Manzungu, E., T. Mpho & A. Mpale-Mudanga (2009). "Continuing discontinuities: local and state perspectives on cattle production and water management in Botswana". *Water Alternatives* 2(2): 205–24.

Marr, S. (2019). "'A town new and modern in conception': non-racial dreams and racial realities in the making of Gaborone, Botswana". *Social Identities* 25(1): 41–57.

Marx, K. (1926) [1818–83]. *The Essentials of Marx; The Communist Manifesto, by Karl Marx and Frederick Engels; Wage-labor and Capital; Value, Price and Profit, and Other Selections, by Karl Marx*. New York, NY: Vanguard Press.

Mathias, S., S. Reaney & P. Kenabatho (2021). "Transmission loss estimation for ephemeral sand rivers in Southern Africa". *Journal of Hydrology* 600: 126487.

Mawdsley, E. (2015). "DFID, the private sector and the re-centering of an economic growth agenda in international development". *Global Society* 29(3): 339–58.

Mbaiwa, J. (2017a). "Ecotourism in Botswana: 30 years later". In K. Backman & I. Munanura (eds), *Ecotourism in Sub-Saharan Africa*, 110–28. London: Routledge.

Mbaiwa, J. (2017b). "Poverty or riches: who benefits from the booming tourism industry in Botswana?" *Journal of Contemporary African Studies* 35(1): 93–112.

McCann, J. (2005). *Maize and Grace: Africa's Encounter with a New World Crop, 1500–2000*. Cambridge, MA: Harvard University Press.

McCusker, B., W. Moseley & M. Ramutsindela (2016a). "Tenure reform and small-scale agriculture in the 'Coloured Reserves'". In B. McCusker, W. Moseley & M. Ramutsindela, *Land Reform in South Africa: An Uneven Transformation*, 139–52. Lanham, MD: Rowman & Littlefield.

McCusker, B., W. Moseley & M. Ramutsindela (2016b). *Land Reform in South Africa: An Uneven Transformation*. Lanham, MD: Rowman & Littlefield.

McIntosh, R. (2005). *Ancient Middle Niger: Urbanism and the Self-Organizing Landscape*. Cambridge: Cambridge University Press.

McKeon, N. (2020). "'Getting to the root causes of migration' in West Africa – whose history, framing and agency counts?" In C. Schierup *et al.* (eds), *Migration, Civil Society and Global Governance*, 140–55. Abingdon: Routledge.

McKeon, N., M. Watts & W. Wolford (2004). *Peasant Associations in Theory and Practice*. Geneva: UNRISD.

McMillan, D. (1995). *Sahel Visions: Planned Settlement and River Blindness Control in Burkina Faso*. Tuscon, AZ: University of Arizona Press.

Meadows, D. *et al.* (2018). "The limits to growth". In G. Dabelko & K. Conca (eds), *Green Planet Blues*, 25–9. New York, NY: Routledge.

Meadows, M. (2000). "The ecological resource base: biodiversity and conservation". In R. Fox & K. Rowntree (eds), *The Geography of South Africa in a Changing World*, 361–89. Oxford: Oxford University Press.

Meloni, G. & J. Swinnen (2018). "Trade and terroir: the political economy of the world's first geographical indications". *Food Policy* 81: 1–20.

Merten, M. (2022). "Controversial Expropriation Bill is finally approved after navigating a 14-year rocky road". *Daily Maverick*, 29 September. https://www.dailymaverick.co.za/article/2022-09-29-controversial-expropriation-bill-is-finally-approved-after-navigating-a-14-year-rocky-road/.

Mesnage, R. *et al.* (2015). "Potential toxic effects of glyphosate and its commercial formulations below regulatory limits". *Food and Chemical Toxicology* 84: 133–53.

REFERENCES

Mogalakwe, M. & F. Nyamnjoh (2017). "Botswana at 50: democratic deficit, elite corruption and poverty in the midst of plenty". *Journal of Contemporary African Studies* 35(1): 1–14.

Montenegro de Wit, M. *et al.* (2021). "Resetting power in global food governance: the UN Food Systems Summit". *Development* 64: 153–61.

Moorehead, R. (1991). *Structural Chaos: Community and State Management of Common Property in Mali.* PhD Dissertation, Institute for Development Studies, University of Sussex.

Morgan, J. (2018). *Correcting for the Inconveniences of Cultivation: Foraging as a Food Source in Southwestern Burkina Faso.* Geography Honors Projects, Macalester College. https://digitalcommons.macalester.edu/geography_honors/55.

Morgan, J. & W. Moseley (2020). "The secret is in the sauce: foraged food and dietary diversity among female farmers in southwestern Burkina Faso". *Canadian Journal of Development Studies/Revue canadienne d'études du développement* 41(2): 296–313.

Morwaeng, L. (2023). "Botswana: NSC pipeline completion quenches thirst in Molepolole". *Daily News*, 15 October. https://allafrica.com/stories/202310160106.html.

Moseley, W. (1993). *Indigenous Agroecological Knowledge among the Bambara of Djitoumou, Mali.* Master's Thesis, School of Natural Resources and Environment, University of Michigan, Ann Arbor. https://deepblue.lib.umich.edu/handle/2027.42/114616.

Moseley, W. (1995). "Securing livelihoods in marginal environments: can NGOs make a long term difference?" In *Policy In The Making, Poverty and Food Economy: Assessing Livelihoods*, 12–22. London: Save the Children Fund. Policy Development Unit Discussion Paper No. 4.

Moseley, W. (2000). *Paradoxical Constraints to Agricultural Intensification in Malawi: The Interplay between Labor, Land and Policy.* Department of Geography, Discussion Paper Series No. 00-1, University of Georgia, Athens.

Moseley, W. (2001). *Sahelian "White Gold" and Rural Poverty–Environment Interactions: The Political Ecology of Cotton Production, Environmental Change, and Household Food Economy in Mali.* PhD dissertation, Department of Geography, University of Georgia. https://www.proquest.com/openview/eea261963569dd2a906f98a14aeb1606/1?pq-origsite=gscholar&cbl=18750&diss=y.

Moseley, W. (2005a). "Reflecting on *National Geographic Magazine* and academic geography" September 2005 special issue on Africa. *African Geographical Review* 24: 93–100.

Moseley, W. (2005b). "Global cotton and local environmental management: the political ecology of rich and poor small-hold farmers in southern Mali". *Geographical Journal* 171(1): 36–55.

Moseley, W. (2006a). "Farm workers, agricultural transformation and land reform in the Western Cape Province, South Africa". *Focus on Geography* 49(1): 1–7.

Moseley, W. (2006b). "Post-Apartheid vineyards: land redistribution begins to transform South Africa's wine country". *Dollars & Sense Magazine*, Jan/Feb: 16–21.

Moseley, W. (2007a). "Mali". In P. Robbins (ed.), *Encyclopedia of Environment and Society*, Vol. 3, 1085–86. London: Sage.

Moseley, W. (2007b). "Neoliberal agricultural policy versus agrarian justice: farm workers and land redistribution in South Africa's Western Cape Province". *South African Geographical Journal* 89(1): 4–13.

Moseley, W. (2007c). "Transformation and the delinquent South African wine connoisseur". *Op-ed in Cape Argus*, 19 March. https://www.researchgate.net/publication/39730455_Transformation_and_the_Delinquent_South_African_Wine_Connoisseur.

Moseley, W. (2007d). "Collaborating in the field, working for change: reflecting on partnerships between academics, development organizations and rural communities in Africa". *Singapore Journal of Tropical Geography* 28: 334–47.

Moseley, W. (2008a). "Strengthening livelihoods in Sahelian West Africa: the geography of development and underdevelopment in a peripheral Region". *Geographische Rundschau International Edition* 4(4): 44–50.

Moseley, W. (2008b). "Mali's cotton conundrum: commodity production and development on the periphery". In W. Moseley & L. Gray (eds), *Hanging by a Thread: Cotton, Globalization and Poverty in Africa*, 83–102. Athens, OH: Ohio University Press.

Moseley, W. (2008c). "Fair trade wine: South Africa's post apartheid vineyards and the global economy". *Globalizations* 5(2): 291–304.

Moseley, W. (2009). "Response to Michael Watts". In A. Samatar (ed.), *Whither Development? The Struggle for Livelihood in the Time of Globalization. Macalester International* Vol 24 (summer), 140–9. Saint Paul: Macalester College.

Moseley, W. (2011). "Lessons from the 2008 global food crisis: agro-food dynamics in Mali". *Development in Practice* 21(4/5): 604–12.

Moseley, W. (2012a). "The corporate take-over of African food security". *Pambazuka News*, 1 November. https://www.pambazuka.org/food-health/corporate-take-over-african-food-security.

Moseley, W. (2012b). "Africa's future? Botswana's growth with hunger". Al jazeera (English), 8 May. https://www.aljazeera.com/opinions/2012/5/7/africas-future-botswanas-growth-with-hunger.

Moseley, W. (2012c). "Famine myths: five misunderstandings related to the 2011 hunger crisis in the Horn of Africa". *Dollars and Sense* 299: 17–21.

Moseley, W. (2013a). "The evolving global agri-food system and African-Eurasian food flows". *Eurasian Geography and Economics* 54(1): 5–21.

Moseley, W. (2013b). "Too many elephants in African parks?" Al jazeera (English), 21 April. https://www.aljazeera.com/opinions/2013/4/21/too-many-elephants-in-african-parks.

Moseley, W. (2014a). "Artisanal gold mining's curse on West African farming". *Al jazeera (English)*, 9 July. http://www.aljazeera.com/indepth/opinion/2014/07/artisanal-gold-mining-west-afric-20147372739374988.html.

Moseley, W. (2014b). "The limits of new social entrepreneurship". *Al jazeera (English)*, 22 December. https://www.aljazeera.com/opinions/2014/12/22/the-limits-of-new-social-entrepreneurship.

Moseley, W. (2015a). "Africa's entrepreneurial dilemma". *Al jazeera (English)*, 4 August. https://www.aljazeera.com/opinions/2015/8/4/africas-entrepreneurial-dilemma/.

Moseley, W. (2015b). "Regional value chains and productivity enhancement in Africa". In K. Hanson (ed.), *Contemporary Regional Development in Africa*, 181–200. Farnham: Ashgate.

Moseley, W. (2015c). "Climate change adaptation is not just about vulnerable countries". *African Arguments*, 29 December. https://africanarguments.org/2015/12/climate-change-adaptation-not-just-about-vulnerable-countries/.

Moseley, W. (2015d). "Food security and the 'Green Revolution'". In J. Wright (ed.), *International Encyclopedia of Social and Behavioral Sciences*, 307–10. Amsterdam: Elsevier.

Moseley, W. (2016). "Agriculture on the brink: climate change, labor and smallholder farming in Botswana". *Land* 5(3): 21.

Moseley, W. (2017a). "The new green revolution for Africa: a political ecology critique". *Brown Journal of World Affairs* 23: 177–90.

Moseley, W. (2017b). "One step forward, two steps back in farmer knowledge exchange: 'scaling-up' as Fordist replication in drag". In J. Sumberg (ed.), *Agronomy for Development: The Politics of Knowledge in Agricultural Research*, 79–90. Abingdon: Routledge.

Moseley, W. (2017c). "The minimalist state and donor landscapes: livelihood security in Mali during and after the 2012–2013 coup and rebellion". *African Studies Review* 60(1): 37–51.

Moseley, W. (2018). "Why it's important to recognize multiple food systems in Africa". *The Conversation*, 18 June. https://theconversation.com/why-its-important-to-recognise-multiple-food-systems-in-africa-97134.

Moseley, W. (2021a). "Political agronomy 101: an introduction to the political ecology of industrial cropping systems". In A. Gasparatos & A. Ahmed (eds), *The Political Ecology of Industrial Crops*, 25–44. London: Earthscan/Routledge.

Moseley, W. (2021b). "Inclusive science must follow the UN Food Systems Summit". *Al jazeera (English)*, 28 September. https://www.aljazeera.com/opinions/2021/9/28/to-end-hunger-we-must-return-to-participatory-science.

Moseley, W. (2022a). "Like a bad rain year: the consequences of Russia's invasion of Ukraine for African food security and the need for greater food sovereignty". *Africa is a Country*, 29 June. https://africasacountry.com/2022/06/like-a-bad-rain-year.

Moseley, W. (2022b). "The trouble with drought as an explanation for famine in the Horn and Sahel of Africa". *The Conversation Africa*, 15 February. https://theconversation.com/the-trouble-with-drought-as-an-explanation-for-famine-in-the-horn-and-sahel-of-africa-177071.

Moseley, W. (2022c). "Development assistance and Boserupian intensification under geopolitical isolation: the political ecology of a crop-livestock integration project in Burundi". *Geoforum* 128: 276–85.

Moseley, W. & J. Battersby (2020). "The vulnerability and resilience of African food systems, food security, and nutrition in the context of the COVID-19 pandemic". *African Studies Review* 63(3): 449–61.

Moseley, W., J. Carney & L. Becker (2010). "Neoliberal policy, rural livelihoods and urban food security in West Africa: a comparative study of the Gambia, Côte d'Ivoire and

Mali". *Proceedings of the National Academy of Sciences of the United States of America* 107(13): 5774–9.

Moseley, W., J. Earl & L. Diarra (2002). "La decentralization et les conflits entre agriculteurs et éleveurs dans le delta intérieur du Niger". In D. Orange *et al.* (eds), *La Gestion Intégrée des Ressources Naturelles en Zones Inondables Tropicales.* Collection Colloques et Séminaires, 101–18. Paris: Institut de Recherche pour le Développement (IRD).

Moseley, W. & R. Fehr (2016). "Female farmers suffer most in southern Africa drought". *Al jazeera (English)*, 30 September. https://www.aljazeera.com/opinions/2016/9/30/female-farmers-suffer-most-in-southern-africa-drought.

Moseley, W. & L. Gray (eds) (2008). *Hanging by a Thread: Cotton, Globalization and Poverty in Africa*. Athens, OH: Ohio University Press.

Moseley, W. & B. Hoffman (2017). "Introduction: hope, despair and the future of Mali". *African Studies Review* 60(1): 5–14.

Moseley, W. & P. Laris (2008). "West African environmental narratives and development-volunteer praxis". *Geographical Review* 98(1): 59–81.

Moseley, W. & B. Logan (2001). "Conceptualizing hunger dynamics: a critical examination of two famine early warning methodologies in Zimbabwe". *Applied Geography* 21(3): 223–48.

Moseley, W. & B. Logan (2005). "Food security". In B. Wisner, C. Toulmin & R. Chitiga (eds), *Toward a New Map of Africa*, 133–52. London: Earthscan.

Moseley, W. & B. McCusker (2008). "Fighting fire with a broken tea cup: a comparative analysis of South Africa's land redistribution program". *Geographical Review* 98(3): 322–38.

Moseley, W. & K. Otiso (eds) (2022). *Debating African Issues: Conversations Under the Palaver Tree*. Abingdon: Routledge.

Moseley, W. & M. Ouedraogo (2022). "When agronomy flirts with markets, gender, and nutrition: a political ecology of the New Green Revolution for Africa and women's food security in Burkina Faso". *African Studies Review* 65(1): 41–65.

Moseley, W. & E. Pessereau (2022). "Mother's little helper: a feminist political ecology of West Africa's herbicide revolution". In H.R. Barcus, R. Jones & S. Schmitz (eds), *Rural Transformations: Globalization and its Implications for Rural People, Land, and Economies*, Pp. 151–66. London: Taylor & Francis Group.

Moseley, W. *et al.* (2013). *An Introduction to Human-Environment Geography: Local Dynamics and Global Processes*. Hoboken, NJ: Wiley/Blackwell.

Moseley, W., M. Schnurr & R. Bezner Kerr (2015). "Interrogating the technocratic (neoliberal) agenda for agricultural development and hunger alleviation in Africa". *African Geographical Review* 34(1): 1–7.

Moyo, D. (2009). *Dead Aid: Why Aid Is Not Working and How There Is a Better Way for Africa*. New York, NY: Farar, Straus & Giroux.

Munro, W. & R. Schurman (2022). "Building an ideational and institutional architecture for Africa's agricultural transformation". *African Studies Review* 65(1): 16–40.

Napier-Bax, P. & D. Sheldrick (1963). "Some preliminary observations on the food of elephant in the Tsavo National Park (East) of Kenya". *East African Wildlife Journal* 1: 40–54.

Ndhleve, S. *et al.* (2021). "Household food insecurity status and determinants: the case of Botswana and South Africa". *AGRARIS: Journal of Agribusiness and Rural Development Research* 7(2): 207–24.

Ndlovu-Gatsheni, S. (2018). "Rhodes must fall". In S. Ndlovu-Gatsheni (ed.), *Epistemic Freedom in Africa: Deprovincialization and Decolonization*, 221–42. Abingdon: Routledge.

NEPAD/CAADP (2023). *"Introducing the Comprehensive Africa Agriculture Development Programme"*. file:///C:/Users/moseley/Downloads/Introducing%20CAADP_English. pdf.

NimbleFins (2023). "Average UK household cost of food 2023". https://www.nimblefins. co.uk/average-uk-household-cost-food.

Nnyepi, M. *et al.* (2015). "Evidence of nutrition transition in Southern Africa". *Proceedings of the Nutrition Society* 74(4): 478–86.

Nyantakyi-Frimpong, H. (2017). "Agricultural diversification and dietary diversity: a feminist political ecology of the everyday experiences of landless and smallholder households in northern Ghana". *Geoforum* 86: 63–75.

Nyantakyi-Frimpong, H. & R. Bezner Kerr (2015). "A political ecology of high-input agriculture in northern Ghana". *African Geographical Review* 34(1): 13–35.

Nyantakyi-Frimpong, H. *et al.* (2017). "Agroecology and healthy food systems in semi-humid tropical Africa: participatory research with vulnerable farming households in Malawi". *Acta Tropica* 175: 42–9.

Nyéléni Declaration (2007). *Declaration of NyNyéléléni, Sélingué, Mali.* https://nyeleni. org/IMG/pdf/DeclNyeleni-en.pdf.

Page, G. (2012). "How to ensure the world's food supply". *Washington Post*, 2 August. https://www.washingtonpost.com/opinions/how-to-ensure-the-worlds-food-supply/2012/08/02/gJQANPGQSX_story.html.

Painter, D. (1978). "Malian socialism reconsidered" Review of *Planning and Economic Policy: Socialist Mali and Her Neighbors; Tradition and Progress in the African Village: Non-Capitalist Transformation of Rural Communities* in Mali, by W. I. Jones & K. Ernst]. *ASA Review of Books* 4: 75–8.

Palchick, M. (2008). *Agricultural Transformation and Livelihood Struggles in South Africa's Western Cape.* Geography Honors Projects, Macalester College. https:// digitalcommons.macalester.edu/geography_honors/13.

Parsons, N. (1998). *King Khama, Emperor Joe and the Great White Queen: Victorian Britain through African Eyes.* Chicago, IL: University of Chicago Press.

Parsons, N. (2023). "Botswana". *Encyclopedia Britannica.* https://www.britannica.com/ place/Botswana.

Patel, R. (2013). "The long green revolution". *Journal of Peasant Studies* 40(1): 1–63.

Patel, R. & P. McMichael (2014). "A political economy of the food riot". In D. Pritchard (ed.), *Riot, Unrest and Protest on the Global Stage*, 237–61. London: Palgrave Macmillan.

Peet, R. (2002a). "Neoliberalism in South Africa". In B. Logan (ed.), *Globalization, the Third World State and Poverty-Alleviation in The Twenty-First Century*, 125–41. Aldershot: Ashgate.

Peet, R. (2002b). "Ideology, discourse, and the geography of hegemony: from socialist to neoliberal development in post apartheid South Africa". *Antipode* 34(1): 54–84.

Peet, R. & M. Watts (eds) (2002a). *Liberation Ecologies: Environment, Development, Social Movements*. London: Routledge.

Peet, R. & M. Watts (2002b). "Liberation ecology: development, sustainability, and environment in an age of market triumphalism". In R. Peet & M. Watts (eds), *Liberation Ecologies*, 1–45. London: Routledge.

Perfecto, I. & J. Vandermeer (2010). "The agroecological matrix as alternative to the land-sparing/agriculture intensification model". *Proceedings of the National Academy of Sciences* 107(13): 5786–5791.

Peters, P. (1984). "Struggles over water, struggles over meaning: cattle, water and the state in Botswana". *Africa* 54(3): 29–49.

Peters, P. (1994). *Dividing the Commons: Politics, Policy, and Culture in Botswana*. Charlottesville, VA: University Press of Virginia.

Peterson, B. (2008). "History, memory and the legacy of Samori in southern Mali, c. 1880–1898". *Journal of African History* 49(2): 261–79.

Peyton, S., W. Moseley & J. Battersby (2015). "Implications of supermarket expansion on urban food security in Cape Town, South Africa". *African Geographical Review* 34(1): 36–54.

Place, F. *et al.* (2022). *"Agroecologically-conducive policies: a review of recent advances and remaining challenges"*. Working Paper #1. Transformative Partnership Program on Agroecology.

Pokorny, B. *et al.* (2019). "All the gold for nothing? Impacts of mining on rural livelihoods in Northern Burkina Faso". *World Development* 119: 23–39.

Popkin, B. (2004). "The nutrition transition: an overview of world patterns of change". *Nutrition reviews* 62(suppl 2): S140–43.

Pretty, J. (2006). *Agroecological Approaches to Agricultural Development*. Rimisp-Latin American Center for Rural Development. Background paper for the World Development Report 2008. https://openknowledge.worldbank.org/server/api/core/bitstreams/32db0802-455a-573a-a135-1b4af4e0cc9c/content.

Quinn, K. (2014). "A tribute to Norman Borlaug". *Crop Science Society of America (CSA) News* 59(11): 20–21.

Ravillion, M. & S. Chen (2007). "China's (uneven) progress against poverty". *Journal of Development Economics* 82: 1–42.

Reardon, T. & R. Hopkins (2006). "The supermarket revolution in developing countries: policies to address emerging tensions among supermarkets, suppliers and traditional retailers". *European Journal of Development Research* 18(4): 522–45.

Republic of South Africa (RSA) (2019). *Report by the Presidential Advisory Panel on Land Reform and Agriculture*. https://www.gov.za/sites/default/files/gcis_document/201907/panelreportlandreform_0.pdf.

Ricardo, D. (1951) [1817]. "On the principles of political economy and taxation". In *Works and Correspondence of David Ricardo*, Volume I, edited by P. Sraffa. Cambridge: Cambridge University Press.

Richards, P. (1985). *Indigenous Agricultural Revolution: Ecology and Food Production in West Africa*. London: Routledge.

Richmond, M. (2018). "Glyphosate: a review of its global use, environmental impact, and potential health effects on humans and other species". *Journal of Environmental Studies and Sciences* 8: 416–34.

Robbins, P. (2004). *Political Ecology: A Critical Introduction*. Oxford: Blackwell.

Roberge, P. (2002). "Afrikaans: considering origins". In R. Mesthrie (ed.), *Language in South Africa*, 79–103. Cambridge: Cambridge University Press.

Roberts, R. (1996). *Two Worlds of Cotton: Colonialism and the Regional Economy in the French Soudan, 1800–1946*. Stanford, CA: Stanford University Press.

Robinson C. (2000). *Black Marxism and the Making of the Black Radical Tradition*. North Carolina, NC: University of North Carolina Press.

Rocheleau, D., B. Thomas-Slayter & E. Wangari (2013). *Feminist Political Ecology: Global Issues and Local Experience*. Abingdon: Routledge.

Rosenberg, D. (2011). "Food and the Arab spring". Gloria Center, 27.

Ross, C. (2014). "The plantation paradigm: colonial agronomy, African farmers, and the global cocoa boom, 1870s–1940s". *Journal of Global History* 9(1): 49–71.

Rostow, W. (1960). *The Stages of Economic Growth: A Non-Communist Manifesto*. Cambridge: Cambridge University Press.

Rotberg, R. (1988). *The Founder: Cecil Rhodes and the Pursuit of Power*. Oxford: Oxford University Press.

Roy, A. (2010). "Peasant struggles in Mali: from defending cotton producers' interests to becoming part of the Malian power structures". *Review of African Political Economy* 37(125): 299–314.

Ruppe, L. (1985). "From the director". *Peace Corps Times*. May/June Issue.

Rusenga, C. (2022). "Rethinking land reform and its contribution to livelihoods in South Africa". *Africa Review* 14(2): 125–50.

Salvato, N. (2022). *Bloodshed, Baptism, Beer: Racial Capitalism and Settler Colonialism on the Medieval Baltic*. Geography Honors Projects, Macalester College. https://digitalcommons.macalester.edu/geography_honors/73.

Samatar, A. (1999). *An African Miracle: State and Class Leadership, and Colonial Legacy in Botswana's Development*. Portsmouth, NH: Heinemann.

Şaul, M. & P. Royer (2001). *West African Challenge to Empire: Culture and History in the Volta-Bani Anticolonial War*. Athens, OH: Ohio University Press.

Schreyger, E. (1984). *L'Office du Niger au Mali, 1932 à 1982: la Problematique d'une Grande Entreprise Agricole dans la Zone du Sahel*. Wiesbaden: Steiner.

Schroeder, R. (1999). *Shady Practices: Agroforestry and Gender Politics in the Gambia*. Berkeley, CA: University of California Press.

Schumacher, E. (1973). *Small Is Beautiful: Economics as if People Mattered*. London: Blond & Briggs.

Schurman, R. (2018). "Micro (soft) managing a 'green revolution' for Africa: the new donor culture and international agricultural development". *World Development* 112: 180–92.

Scully, P. (1992). "Liquor and labor in the Western Cape, 1870–1900". In J. Crush. & C. Ambler (eds), *Liquor and Labor in Southern Africa*, 56–77. Athens, OH: Ohio University Press.

Sen, A. (1982). *Poverty and Famines: An Essay on Entitlement and Deprivation*. New York, NY: Oxford University Press.

Servin, J. & W. Moseley (2023). "The hidden safety net: wild and semi-wild plant consumption and nutrition among women farmers in Southwestern Burkina Faso". *African Geographical Review* 42(4): 483–503.

Sihlobo, W. & J. Kirsten (2021). "Agriculture in South Africa". In A. Oqubay, F. Tregenna & I. Valodia (eds), *Oxford Handbook of the South African Economy*, 195. Oxford: Oxford University Press.

Simplemaps (2023). "World cities database". Simplemaps Interactive Maps and Data. https://simplemaps.com/data/world-cities (accessed 16 May 2024).

Slavchevska, V., S. Kaaria & S. Taivalmaa (2019). "The feminization of agriculture". In T. Allan *et al.* (eds), *Oxford Handbook of Food, Water and Society*, 268. Oxford: Oxford University Press.

Socioeconomic Data and Applications Center (SEDAC) (2010). "Global roads open access data set (gROADS), v1 (1980–2010)". NASA Earthdata. https://sedac.ciesin.columbia.edu/data/set/groads-global-roads-open-access-v1/data-download (accessed 16 May 2024).

South African Wine Industry Information and Systems (SAWIS) (2006). *South African Wine Industry Statistics 2006.* https://www.sawis.co.za/info/annualpublication.php.

Special Rapporteur on the Right to Food (2012). "2008: para. 17"; quoted in *Special Rapporteur on the Right to Food 2012.* https://www.ohchr.org/en/special-procedures/sr-food/about-right-food-and-human-rights.

Stanford University (2013). *Inland waters, Botswana, 2013.* Stanford University Libraries EarthWorks. https://earthworks.stanford.edu/catalog/stanford-fq566jm3042 (accessed 16 May 2024).

Statistics Botswana (2020). *Annual Agricultural Survey Report 2019: Traditional Sector.* Gaborone: Agricultural Statistics Section. https://www.statsbots.org.bw/sites/default/files/publications/ANNUAL%20AGRICULTURAL%20SURVEY%20REPORT%202019%20TRADITIONAL%20SECTOR.pdf.

Stats SA (2022). *Statistical Release P0302.* Pretoria: Department of Statistics South Africa. https://www.statssa.gov.za/publications/P0302/P03022022.pdf.

Stone, G. (2018). "Agriculture as spectacle". *Journal of Political Ecology* 25(1): 656–85.

Sumberg, J. (ed.) (2017). *Agronomy for Development: The Politics of Knowledge in Agricultural Research*. Abingdon: Routledge.

Sumberg, J., J. Thomson & P. Woodhouse (2012). "Why agronomy in the developing world has become contentious". *Agriculture and Human Values* 30(1): 71–83.

Sunday Standard (2013). "Backyard gardens going through a rough patch". *Sunday Standard*, 25 July. https://www.sundaystandard.info/backyard-gardens-going-through-a-rough-patch/.

Swift, J. (1996). "Desertification: narratives, winners and losers". In M. Leach & R. Mearns (eds), *The Lie of the Land: Challenging Received Wisdom on the African Environment*, 73–90. Oxford: James Curry.

Swindale, A. & P. Bilinsky (2006). *Household Dietary Diversity Score (HDDS) for Measurement of Household Food Access: Indicator Guide (v.2)*. Washington, DC: FHI 360/FANTA.

Tamariz, G. & M. Baumann (2022). "Agrobiodiversity change in violent conflict and post-conflict landscapes". *Geoforum* 128: 217–22.

Tefft, J. (2004). *Mali's White Revolution: Smallholder Cotton from 1960 to 2003*. 2020 Vision Briefs 12 No. 5. Washington, DC: International Food Policy Research Institute (IFPRI).

Tefft, J. *et al.* (2000). *Linkages between Agricultural Growth and Improved Child Nutrition in Mali*. Michigan State University International Development Working Papers No. 1096-2016-88422. https://ageconsearch.umn.edu/nanna/record/54575/files/idwp79.pdf?withWatermark=0&withMetadata=0&version=1&registerDownload=1.

Terborgh, J. & C. Van Schaik (2002). "Why the world needs parks". In J. Terborgh *et al.* (eds) *Making Parks Work: Strategies for Preserving Nature*, 3–14. Washington, DC: Island Press.

Termote, C. *et al.* (2022). "Nutrient composition of Parkia biglobosa pulp, raw and fermented seeds: a systematic review". *Critical Reviews in Food Science and Nutrition* 62(1): 119–44.

Theriault, V. & J. Sterns (2012). "The evolution of institutions in the Malian cotton sector: an application of John R. Commons' ideas". *Journal of Economic Issues* 46(4): 941–66.

Thornton, J. (1977). "Demography and history in the Kingdom of Kongo, 1550–1750". *Journal of African History* 18(4): 507–30.

Tkaczyk, Z. & W. Moseley (2023). "Dietary power and self-determination among female farmers in Burkina Faso: a proposal for a food consumption agency metric". *Land* 12(5): 978.

Tlhankane, M & M. Mguni (2023). "Poverty eradication projects flop". *Mmegi Online*, 24 February. https://www.pressreader.com/botswana/mmegi/20230224/281556590016741.

Toledo, V. & P. Moguel (2012). "Coffee and sustainability: the multiple values of traditional shaded coffee". *Journal of Sustainable Agriculture* 36(3): 353–77.

Tondel, F. *et al.* (2020). "Rice trade and value chain development in West Africa". *European Centre for Development Policy Management (ECDPM) and the Initiative prospective agricole et rurale (IPAR)*. Discussion Paper 283. Maastricht, Netherlands. https://www.researchgate.net/profile/Cecilia-Dalessandro/publication/354659090_Rice_trade_and_value_chain_development_in_West_Africa_an_approach_for_more_coherent_policies/links/6144c354a3df59440b946574/Rice-trade-and-value-chain-development-in-West-Africa-an-approach-for-more-coherent-policies.pdf.

Torres Solís, J. & K. Moroka (2011). "Innovative corporate social responsibility in Botswana: the Debswana mining company study case". *Contaduría y administración* 233: 91–104.

Tripp, R. (2001). *Seed Provision and Agricultural Development*. Oxford: James Curry.

Turner, B. (1989). "The specialist-synthesis approach to the revival of geography: the case of cultural ecology". *Annals of the Association of American Geographers* 79(1): 88–100.

United Nations Development Programme (UNDP) (2022). *Human Development Report 2021–22: Uncertain Times, Unsettled Lives: Shaping Our Future in a Transforming World*. https://hdr.undp.org/content/human-development-report-2021-22.

UNDP (2023). *Multidimensional Poverty Index 2023: Country Briefing Botswana*. https://hdr.undp.org/sites/default/files/Country-Profiles/MPI/BWA.pdf.

UNHCR (2017). *Burundi Situation: Displacement of Burundians into Neighbouring Countries*. https://reliefweb.int/map/burundi/burundi-situation-displacement-burundians-neighbouring-countries-april-2015-30-april.

United States Department of Agriculture (USDA) (2023). *Cotton Production Statistics*. https://fas.usda.gov/data/commodities/cotton (accessed 1 July 2023).

University of Missouri (2020). *Watermelon: A Brief History*. Division of Plant Sciences. https://ipm.missouri.edu/MEG/2020/7/watermelon-DT/.

University of Texas (1979). *Perry-Castañeda Library Map Collection: South Africa maps*. University of Texas Libraries. https://maps.lib.utexas.edu/maps/south_africa.html (accessed 16 May 2024).

van Beusekom, M. (1990). *Colonial Rural Development: French Policy and African Response at the Office du Niger, Soudan Français (Mali), 1920–1960*. Dissertation, Johns Hopkins University.

van Beusekom, M. (2000). "Disjunctures in theory and practice: making sense of change in agricultural development at the Office du Niger, 1920–60". *Journal of African History* 41(1): 79–99.

Van der Kooij, S. *et al.* (2013). "The efficiency of drip irrigation unpacked". *Agricultural Water Management* 123: 103–10.

Vandermeer, J. & I. Perfecto (2007). "The agricultural matrix and a future paradigm for conservation". *Conservation Biology* 21(1): 274–7.

Vanloquerin, G. & P. Baret (2009). "How agricultural research systems shape a techno-logical regime that develops genetic engineering but locks out agroecological innova-tions". *Research Policy* 38(6): 971–83.

Von Braun, J. *et al.* (2023). *Science for Transformation of Food Systems: Opportunities for the UN Food Systems Summit*. Cham, CH: Springer.

Von Braun, J. & M. Kalkuhl (2015). *"International science and policy interaction for improved food and nutrition security: toward an International Panel on Food and Nutrition (IPFN)"*. ZEF Working Paper Series No. 142.

Wa Thiong'o, N. (1998). "Decolonising the mind". *Diogenes* 46(184): 101–104.

Waldman, L. (2006). "Klaar gesnap as kleurling: the attempted making and remaking of the Griqua people". *African Studies* 65(2): 175–200.

Waldman, L. (2007). *The Griqua Conundrum: Political and Socio-Cultural Identity in the Northern Cape, South Africa*. New York, NY: Peter Lang.

Walker, D. *et al.* (2018). "Alluvial aquifer characterisation and resource assessment of the Molototsi sand river, Limpopo, South Africa". *Journal of Hydrology: Regional Studies* 19: 177–92.

Watts, M. (1983). *Silent Violence: Food, Famine, and Peasantry in Northern Nigeria*. Berkeley, CA: University of California Press.

Watts, M. (2009). "Oil, development, and the politics of the bottom billion". In A. Samatar (ed.), *Whither Development? The Struggle for Livelihood in the Time of Globalization. Macalester International* Vol 24 (summer), 79–130. Saint Paul: Macalester College.

Werthmann, K. (2009). "Working in a boom-town: female perspectives on gold-mining in Burkina Faso". *Resources Policy* 34: 18–23.

Werthmann, K. (2017). "The drawbacks of privatization: artisanal gold mining in Burkina Faso 1986–2016". *Resources Policy* 52: 418–26.

Wezel, A. *et al.* (2009). "Agroecology as a science, a movement and a practice: a review". *Agronomy for Sustainable Development* 29: 503–15.

Widgren, M. (2012). "Slaves: inequality and sustainable agriculture in pre-colonial West Africa". In A. Hornborg, B. Clark & K. Hermele (eds), *Ecology and Power: Struggles over Land and Material Resources in the Past, Present and Future*, 97–107. Abingdon: Routledge.

Widgren, M. (2017). "Agricultural intensification in Sub-Saharan Africa, 1500–1800". *Economic Development and Environmental History in the Anthropocene: Perspectives on Asia and Africa*, 51–67. London: Bloomsbury.

Wikipedia (2023). "Think globally, act locally". https://en.wikipedia.org/wiki/Think_globally,_act_locally (accessed 7 November 2023).

Williams, G. (2005). "Black economic empowerment in the South African wine industry". *Journal of Agrarian Change* 5(4): 476–504.

Wise, T. (2020). "Africa's choice: Africa's green revolution *has* failed, time to change course". Institute for Agriculture and Trade Policy. https://www.iatp.org/africas-choice (accessed 15 May 2024).

World Bank (1992). *World Development Report 1992: Development and the Environment*. New York, NY: Oxford University Press.

World Bank (2018). *Africa: Water Bodies*. World Bank Data Catalog. https://datacatalog.worldbank.org/search/dataset/0040797 (accessed 16 May 2024).

World Bank (2023a). *The World Bank in Botswana*. https://www.worldbank.org/en/country/botswana/overview (accessed 18 August 2023).

World Bank (2023b). *World Bank Data Botswana*. https://data.worldbank.org/country/botswana (accessed 18 August 2023).

World Bank (2023c). *World Bank Data: Gini Index*. https://data.worldbank.org/indicator/SI.POV.GINI (accessed 16 May 2024).

World Bank (2023d). *World Bank Data: Urban Population: Botswana*. https://data.worldbank.org/indicator/SP.URB.TOTL.IN.ZS?locations=BW (accessed 21 August 2023).

World Bank (2023e). *Poverty and Equity Brief: Botswana*. https://databankfiles.worldbank.org/public/ddpext_download/poverty/987B9C90-CB9F-4D93-AE8C-750588BF00QA/current/Global_POVEQ_BWA.pdf.

World Bank (2023f). "World maps of the Köppen-Geiger Climate Classification". World Bank Catalog. https://datacatalog.worldbank.org/search/dataset/0042325 (accessed 15 May 2024).

World Bank (2023g). "Burkina Faso roads". World Bank Catalog. https://datacatalog.worldbank.org/search/dataset/0041083/Burkina-Faso-Roads (accessed 15 May 2024).

World Bank (2024). *World Bank Data Mali.* https://data.worldbank.org/country/Mali (accessed 24 February 2024).

World Wildlife Fund (2012). "Terrestrial ecoregions of the world". https://www.worldwildlife.org/publications/terrestrial-ecoregions-of-the-world (accessed 16 May 2024).

Wudil, A. *et al.* (2022). "Reversing years for global food security: a review of the food security situation in sub-Saharan Africa (SSA)". *International Journal of Environmental Research and Public Health* 19(22): 14836.

Zhou, P. *et al.* (2013). "Botswana". In S. Hachigonta *et al.* (eds), *Southern African Agriculture and Climate Change: A Comprehensive Analysis.* Washington, DC: International Food Policy Research Institute.

INDEX